ECKART EHLERS AND THOMAS KRAFFT (EDS.)

SHÂHJAHÂNÂBÂD/OLD DELHI

ERDKUNDLICHES WISSEN

SCHRIFTENREIHE FÜR FORSCHUNG UND PRAXIS
BEGRÜNDET VON EMIL MEYNEN
HERAUSGEGEBEN VON GERD KOHLHEPP
IN VERBINDUNG MIT ADOLF LEIDLMAIR UND FRED SCHOLZ

HEFT 111

FRANZ STEINER VERLAG STUTTGART
1993

ECKART EHLERS AND THOMAS KRAFFT (EDS.)

SHÂHJAHÂNÂBÂD/ OLD DELHI

TRADITION AND COLONIAL CHANGE

FRANZ STEINER VERLAG STUTTGART

1993

Die Deutsche Bibliothek - CIP Einheitsaufnahme
Shâhjahânâbâd, Old Delhi : tradition and colonial change /
Eckart Ehlers and Thomas Krafft (ed.). - Stuttgart : Steiner,
1993
(Erdkundliches Wissen ; H. 111)
ISBN 3-515-06218-1
NE: Ehlers, Eckart [Hrsg.]; GT

 Gedruckt mit Unterstützung der Deutschen Forschungsgemeinschaf.
Druck: Druckerei Proff, Eurasburg.
Printed in Germany

Contents

Preface

The following collection of articles deals with Old Delhi, formerly and traditionally known as Shâhjahânâbâd - a creation of the great Moghul ruler Shâh Jahân (1628-1658). All articles have their origin in an at first more accidental encounter with 19th century city plans of the Indian subcontinent. While searching for documents and materials in regard to the question of the variational breadth of cityscapes in the Islamic world, the undersigned came across a great number of printed and hand-drawn maps of Indian cities in the the Map Archives of the India Office Library in London. Among the many maps, a unique piece of extremely conspicuous cartography stood out: a hand-drawn and hand-coloured map of Shâhjahânâbâd/Old Delhi. Striking not only by its size of approximately 100 by 100 cm, it also contained a wealth of detailed informations in written form or in symbols and colours, more than any in comparison to the published official maps of the British administration. Besides, this map seemed to represent on first sight all those phenomena that hitherto were considered to be constitutive of the "ideal" city of the Islamic world.

Without going into further details at this place, one should mention the fact that this map is to be dated around 1850. Crucial indications for such a dating are the following facts. It contains urban and architectural fabric which definitely has not been completed before the late 1840ies, e. g. Ellenborough's Tank (in the map: Lal Diggi) constructed in 1846 or a pavillion (Zafar Mahall) in the Emperor's Garden of the Red Fort, built in 1847. On the other hand the map represents definitely pre-mutiny Shâhjahânâbâd , i.e. the city before 1857.

Critical and lengthy examination of this map revealed that it seems to have been known for some time. Both Susan Gole (London) and Attilio Petruccioli (Rome) mention or describe it in some detail in publications (quoted in the following articles), but never in respect to their importance for the urban geography and history of the Islamic world.

The editors of this volume are greatly indebted to Dr. Andrew S. Cook, the head of the Map Section of the Oriental and India Office collections of the British Museum. He not only provided all help and assistance in the early stages of the "on site work" with the map, but was likewise instrumental and supportive in regard to our different wishes of photographic reproductions of the whole map or parts of it. The fact that in the meantime the originally damaged and somewhat dilapidated map has been restored in the laboratories of the British Museum is also due to his keen and engaged interest in this map and the whole collection. Dr. Cook also helped in obtaining the right to reproduce this map!

The close and continued work with this unique map made it more and more desirable to make it available to a broader public. After long and careful discussions with the cartographers of the Department of Geography of the Rheinische Friedrich

Wilhelms-Universität Bonn, it was decided to dare a reconstruction and—first of all—a redrawing of the whole map in its original size. Mr. Gerd Storbeck, engineer of cartography in the Department of Geography of the University of Bonn, assumed the likewise difficult and delicate task of redrawing the topographic situation and of developing an adequate colouring for the new map. The cartographically problematic state of the original and obvious deficiencies of its photographic reproductions made it necessary to check and examine the draft in London several times. However, the comparison of the original state of this map (see frontispiece) with its redrawing surely proves the justification of all these endeavours. The editors are most grateful to Mr. Storbeck for his painstaking, tenacious and also enthusiastic work.

With approaching completion of the new map it became more and more obvious that at least a partial scientific analysis of the rich contents of this map and of its potentials for scientific discussions would be appropriate. The editors are grateful to Professor Narayani Gupta, historian at the Jamia Milla Islamia University at Delhi, to Dr. Jamal Malik (Heidelberg/Bonn) and to Mrs. Susan Gole (London), author of a well-known book on Indian city maps, for their willingness to contribute to this volume. Special thanks go to Dr. Malik, whose expertise in the economic and social history of 19th and 20th century urbanism in the Indian subcontinent helped to solve many problems with the Persian inscriptions of the map and also in identifying Islamic institutions and their importance for Shahjahanabad's urban culture. Dr. Tilak Raj Chopra's help (Bonn) in transcribing and homogenizing names and locations is gratefully acknowledged.

In spite of only five contributions, the editors agreed to some author's explicit wishes to maintain his or her specific arrangement of their texts and its annotations. This consideration of individual preferences for publication is the reason for a certain variety in the presentation of the articles.

Great thanks are due to the Deutsche Forschungsgemeinschaft for the financial support of the research project on the typological and regional versatility of Islamic urbanism and urbanity. The Deutsche Forschungsgemeinschaft also supported the print of the attached map, which is equally acknowledged with thanks. Finally, thanks are due to Mrs. Beate Zerbe (Bonn) for several versions of some of the manuscripts and Mr. Albrecht W. Kraas (Bonn) for the preparation of the texts for print and the layout of the whole volume.

Bonn, July 1993

Eckart Ehlers
Thomas Krafft

Map of Shāhjahānābād / Delhi in its original status in the Map Section of the Oriental and India Office Collections (Archive Reference: India Office Records X, 1659).

The incomplete colouring of parts of the city refers to the urban differentiation into different wards or thânâs.

The scale of the original map and of the reproduction in this volume is approximately 1: 2800; the scale of this frontispiece is approximately 1: 21000.

Islamic Cities in India? - Theoretical Concepts and the Case of Shâhjahânâbâd/Old Delhi

Eckart Ehlers (Bonn) - Thomas Krafft (Marburg)

The never ending flow of academic speculation about the distinct and unique character of the city of the Islamic Middle East has recently gained new impetus and taken on new directions. A great number of new studies have been published. Some of them ask very distinctly: "What is Islamic about the Islamic City?" (Abu-Lughod 1989). Others offer new answers or new assumptions regarding this very question. Bonine (1990) e.g. points to potentially close interrelationships between "sacred directions and city structure", while Wirth (1991) emphasises the aspect of "privacy" as a distinct and unique feature of Islamic urbanity. On the other hand, serious doubts have been formulated about the whole concept of the "Islamic City" and its "orientalistic" interpretation (Abu-Lughod 1987). Whatever argument, view or interpretation has been brought forward, a few facts can nowadays be considered as constituents if not of the "Islamic City", then surely of the "City of the Islamic Middle East". The question is and remains whether these facts are sufficient for the identification and convincing interpretation of the phenomenon as such.

Theoretical Concepts and Problems

A review of international and interdisciplinary academic literature on the concept of "Islamic cities" reveals a remarkable uniformity in regard to specific forms and functions (Ehlers 1992, 1993). Without going into further detail, it may suffice to refer to figure 1. It shows a collection of city models by German, British, French and Israeli authors—mostly geographers—who have summarized their interpretations of what they consider to be part of, possibly even the very essence of, this controversial subject and topic.

No doubt for many researchers the city of the Islamic Middle East seems to be a kind of formal stereotype. Its constituents prove to be the same again and again: a centrally located Friday Mosque, the bazaar around it with very distinct socio-economic differentiations from centre to periphery, a city wall and citadel, intraurban quarters, blind alleys and similar features.

From a geographical perspective, we need to consider in addition to these purely formal aspects a set of factors which, not in themselves but in combination with each other, may contribute to a better understanding of the uniqueness and very "essence" of Islamic urbanity. As pointed out elsewhere (Ehlers 1992, 1993), these factors include:

- very specific city-hinterland relationships of a rent-capitalist nature,
- spatial, socio-economic and political impacts of institutions such as waqf,

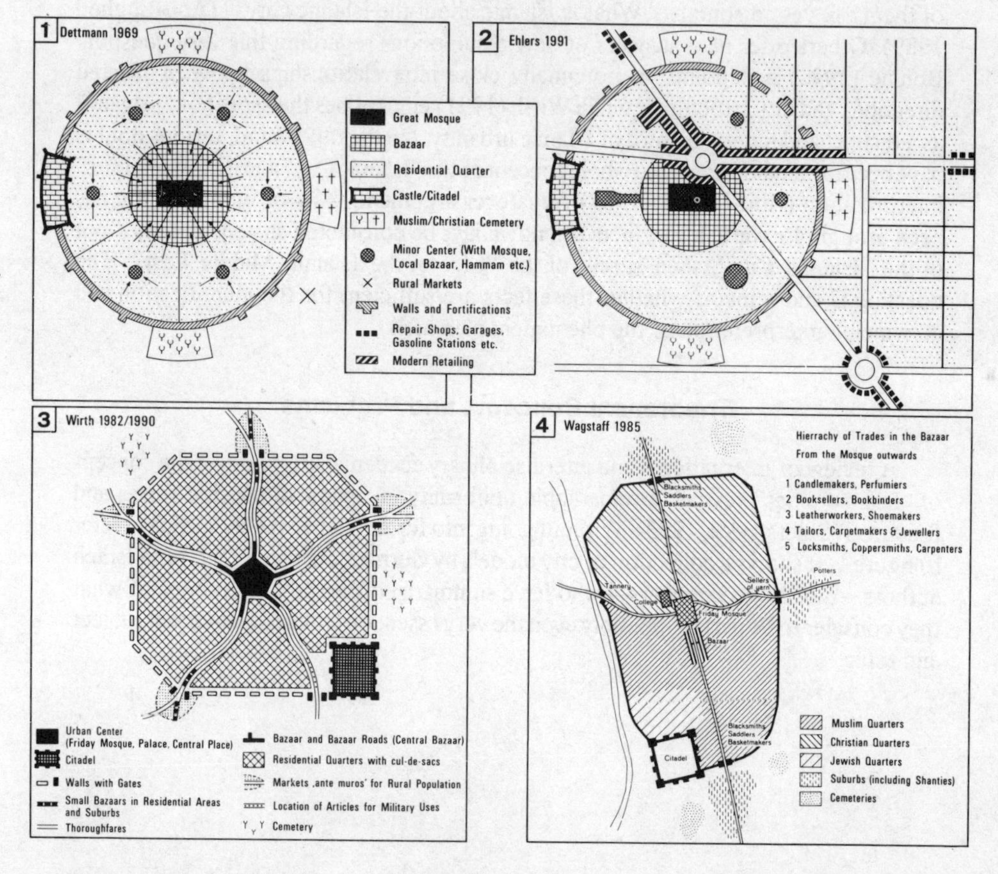

Fig. 1: Models of the "Islamic City" and of the Bazaar (According to Different Authors)

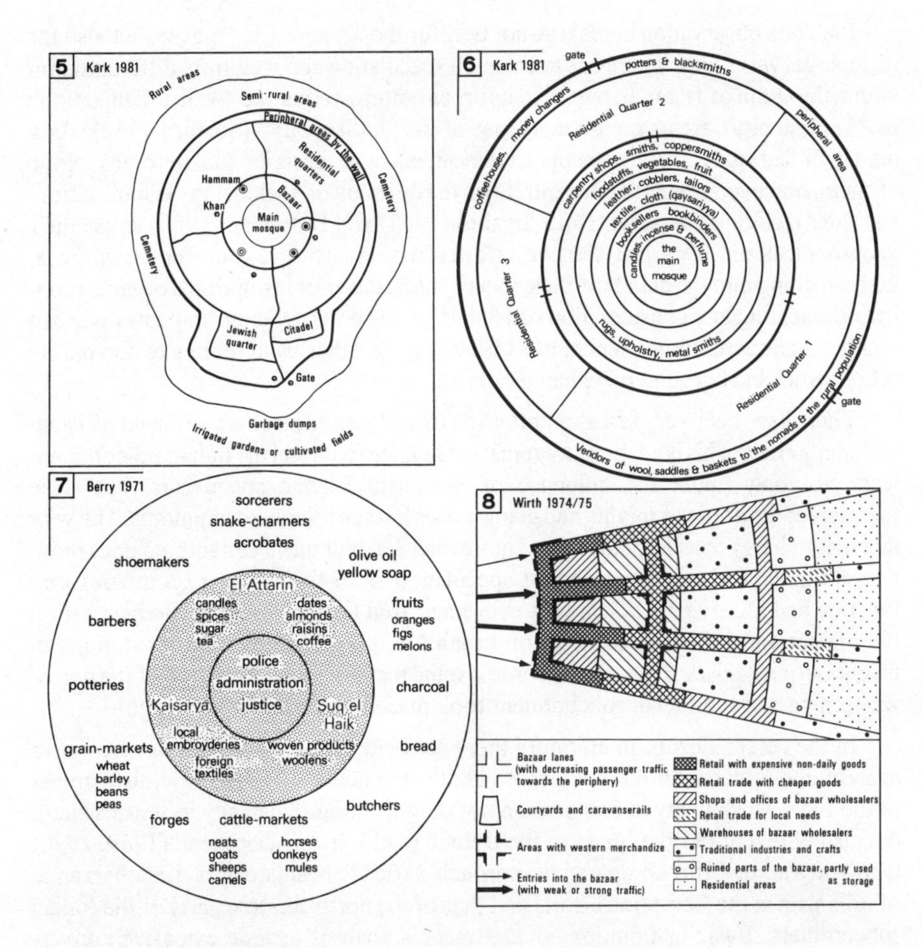

Fig. 1: Models of the "Islamic City" and of the Bazaar (According to Different Authors)

- a historically developed and persisting central place system of dominance.

This list is by no means complete, and further non-geographic factors may be even more important for defining our problem. So far, neither religious historical anthropological nor other disciplinary approaches have been incorporated into the predominantly phenomenological geographical interpretations of the "Islamic City"-concept—with the well-known exception of v. Grunebaum's concept and its consequences (Abu-Lughod 1987; v. Grunebaum 1955). Before this background Inalcik's statement carries a lot of weight: "Anthropologists and geographers will discover 'meaning' only after the necessary 'fieldwork' in the court records of Islamic cities has been done" (1990, p.21).

Inalcik's observation holds true not only for the "Islamic City" per se, but also for its regional variations. There has been much speculation about cultural differentiation within the realm of Islam. In regard to its urban culture, terms like "Arab urbanization" or "Turkish city" are as common as that of the "Indian City" (Smailes 1969). As a matter of fact quite a few attempts have been made to describe Indian towns. None of them, however, seems to concentrate on the question of their being "Islamic cities" and their resulting character. Both Smailes (1969) and Dettmann (1980) stress their geographical and historical division into traditional city - cantonment - civil lines, without going into the details of their social, political and/or institutional organization, thus eliminating any consideration of Islamic or any other religious impacts on urban functions and structures. Schlingloff (1969) on the other hand, points to the purely "Hinduistic" background of Indian urbanism.

There have been very few attempts so far to analyse and discuss the impact of Islam on India's cities. This is all the more remarkable since India and the Indian subcontinent were not only under the influence of powerful Islamic conquerors, but were themselves the origin of mighty and glamourous Islamic rulers and kingdoms. The very fact that India's present population of more than 800 million inhabitants includes more than 90 million Muslims means that speculation as to their impact on urban form, function and life styles is at least not irrevelant. And finally, would anybody question the impact and influence of Islam on urban life, physiognomy, and institutions in neighbouring Pakistan? It has to be borne in mind that the political division of the north-western part of the Indian subcontinent took place less than fifty years ago!

In the recent European literature there seems to be only one serious attempt to analyse and understand the uniqueness and/or the cultural and historic distinctness of the Indo-Muslim city in comparison not only with the Indian city in sensu stricto, but also with that of other parts of the Islamic world. It was Dettmann (1969, 1970) who—in a noteworthy comparative approach—tried to evaluate the variational range of urbanism in the Levantine littoral and that of the north-western parts of the Indian subcontinent. Basic assumptions of Dettmann's analysis include extensive convergence in the overall formal and functional patterns of the cities (Dettmann 1970, pp. 96-102). They include all those elements assembled in figure 1, but also the irregular street pattern and the rent-capitalist economic structure of the cities. On the other hand, differences do exist. They are, however, only of gradual and minor relevance and refer to the following details (Dettmann 1970, pp. 102-109):

> Unlike other cities of the Islamic world the bazaars of the Islamic Indian cities do not have any differentiations. On the contrary: retailing, manufacturing and living form a close symbiosis.
>
> Wholesaling, retailing, manufacturing and services are considered to cover a broader range, partly due to an "untypical" inclusion to be found in the traditional bazaars of Arab, Turkish or Persian cities.

Bazaars are by no means spatially extended complexes, but are characterized by linear patterns,

thus excluding and preventing the development of central-peripheral gradients within the traditional business sections of the cities of north-western India and Pakistan.

Besides these predominantly formal and to a certain degree also functional variations within the economic centre of the traditional city, the bazaar, Dettmann (1970, pp. 107-109) also points out certain differences in regard to architectural design and ethnic/religious/caste differentiation.

On the whole, however, Dettmann comes to the conclusion that urbanism and urbanity in the north-western parts of the Indian subcontinent exhibit "to a great extent...a number of common features which in many respects follow similar principles of spatial order" (ibid., p.123). It is before this background that the following documentation and discussion should be viewed. The main questions to be answered are the following:

1) To what extent do traditional cities in the north-western part of the Indian subcontinent correspond to the general—or better, generally accepted—model of the "Islamic city"?
2) To what extent have traditional cities in the north-western part of the Indian subcontinent developed and maintained Islamic institutions and forms of Islamic urbanity?
3) In concentrating our question on India itself: to what extent are we justified in speaking of "Islamic cities" in the case of those cities that were founded and/or organized as cities of a predominantly Muslim population and are now political and administrative centres of a predominantly Hindu society and population?

Answers to these and related questions may be found in the example of Shâhjahânâbâd/Old Delhi.

Islamic City? - Remarks About a Historical City Plan

Old Delhi or Shâhjahânâbâd is—like its colonial expansion and as its name indicates—a planned city. Founded and created by the Mughal emperor Shâhjahân (1628-1658), it was intended to serve as his capital city. Unlike its modern successor, however, which is the capital of a predominantly Hindu state, Shâhjahânâbâd served as one of several capitals of a Muslim state and ruler.

It is neither the intention nor the purpose of this paper to go into the details of Shâhjahânâbâd's historical development and functions (for detailed analyses see Frykenberg, ed. 1986; Gupta 1981; Krafft 1993 and the other articles in this volume!).

It is, however, an obvious fact that not only Old Delhi's present-day structures, but even more so its mid-19th century appearance (see attached historical plan!) are very highly reminiscent of many other imperial Islamic cities. It may therefore be adequate to recollect rather briefly Shâhjahânâbâd's development and growth as an imperial city, and to comment briefly on the physical ingredients of its mid-19th century and pre-mutiny form and appearance.

The Development of Pre-19th Century Shajahanabad

As already mentioned previously, Shâhjahânâbâd is essentially a planned city. At a young age, Shâhjahân had developed a strong interest in architecture and building (Asher 1992, p.171f.). Although imperial building projects were planned by the collective efforts of a court bureau of architects (Koch 1991, p. 96), the emperor had a direct, creative influence. During the first decade of his reign he put much effort into rebuilding parts of the palaces in the imperial cities of Lahore and Agra. Under Shâhjahân, Mughal architecture reached its zenith, with the Taj Mahal as the outstanding symbol of that period. By the twelfth year of his reign Shâhjahân decided to build a new capital. Several sites were considered and discarded before he finally decided on Delhi. The reasons for this decision were both political and religious. For centuries Delhi had been the natural capital for the various rulers of North India, because of its strategic position. It lay in the fertile plains of the Doab, protected by a ring of hills, and controlled the main access route to the Ganga. Shâhjahânâbâd was to become at least the ninth city in a long succession of famous imperial predecessors. The site for the new capital was on the right bank of the Yamuna, north of Dînpanâh and Ferozâbâd, extending up to a small island-fort which had been built in 1546.

From the earliest time of Muslim rule in India, Delhi had been one of the religious centres. In the seventeenth century it was a major place of pilgrimage, and this religious tradition also influenced Shâhjahân's choice of Delhi. Some of the existing shrines and also the Kalân Masjid (built by the prime minister under Feroz Shâh around the 1380s) were incorporated into the new city. The city of Sher Shâh Sûr (about 1540) was now only some yards to the south of Shâhjahânâbâd's southern gate (the Delhi Gate). Some of the previously existing long-distance roads, like the one to Lahore, or such roads as the one leading out from Feroz Shâh's citadel to his hunting-lodge on the northern ridge became part of the new city's street system.

But by and large Shâhjahânâbâd was a planned city. It was intended to be an imperial city expressing the power and grandeur of the Mughal court. The Red Fort was the first complex of buildings to be constructed. It was located at the north-eastern corner of the city, right on the bank of the Yamuna, where it made full use of the river as a natural defence. In addition the riverside palaces benefitted from the fresh and cool breezes from the east. The fort/palace took almost ten years to complete, and was actually a city within a city. On the 19th of April 1648, the auspicious day selected by the royal astrologers, Shâhjahân entered his new city by the river gate. Surrounding the fort and along the main axes of the city were extensive gardens, elegant palaces and mosques for the royal family and other members of the nobility.

The urban infrastructure was laid out in a rather formal, geometric pattern and shows traces of both Persian and Hindu traditions of town planning and architecture. The Persian influence largely accounts for the formalism and symmetry of the palaces, gardens and boulevards. Undoubtedly Shâhjahân was very well aware of the layout of the imperial Isfahan of Shâh Abbâs. He had sent three envoys to the court in Isfahan seeking military support when he was still a rebel prince, and maintained contact with the Persian empire throughout his reign. It is probably through his envoys, who spent considerable time in the Persian capital, that Shâhjahân got detailed information about the architectural concepts of Isfahan.

The designed infrastructure of Shâhjahânâbâd comprised the fort, the Friday Mosque and the other major mosques, including the corresponding waqf properties, the two main boulevards and the bazaars around the Friday Mosque, the elaborate system of water channels, some of the major gardens, and the surrounding city wall. The arrangement of these planned elements was influenced by certain site features, which preclude absolute symmetry. The rest of the city was left to individual development. Over time Shâhjahân allocated sections of the land arround the fort to important members of the court for their palaces and mansions. The dependents of the nobles in turn built their residences around these mansions, which were the nuclei for the development of the mahallah system.

The city grew rapidly and estimates of the city's population range from 150,000 to almost 500,000 by the end of Shâhjahân's reign. Aurangzeb and his successor spent most of their reign on the move in city-camps in the Deccan and their influence on the morphology of Shâhjahânâbâd was very limited. In the following decades the city was the seat of almost powerless Mughal kings, who saw their city ransacked by various invaders several times. During the second half of the seventeenth century trade and population declined and building activity was reduced to almost nil. Therefore the basic features and structures of the city remained unchanged.

From 1803 to 1857 the East India Company virtually controlled Delhi. The city was taken over by the British after the defeat of the Marathas at the battle of Paṭparganj in 1803. The British Resident, living in an old palace near Kashmiri Gate, represented the company and therefore the actual rulers. The Mughal royal family had to restrict themselves to the position of British "Pensioners". Though the Company referred to the last Mughal Emperor Bahadur Shah as the "King of Delhi", it was quite evident that his empire was actually limited to the area within the walls of his palace. As in other Indian states, the British retained the former rulers as instruments for carrying out British policy and reducing popular resistance. The Revolt of 1857 brought this mockery to a radical end.

During the first phase of British colonial rule in Delhi from 1803 to 1857 the city was still very much a Mughal city. Some transformations occurred, but they were far less dramatic than those that followed after the British had recaptured the city in 1858. The early nineteenth century map of pre-mutiny Delhi presented in this volume is therefore an indespensable source for reconstructing the morphology of Shâhjahân's imperial capital. At the same time, however, it is an important tool for answering the question of its Islamic character.

Morphology and Morphological Elements of 19th Century Shâhjahânâbâd.

Water systems and Canals: The hot, dry climate of Delhi made it necessary to develop an hydraulic strategy in order to ensure a constant, year-round supply of water. Therefore part of the Yamuna had been diverted from a point many miles north of Delhi and coaxed into an interlaced system of channels (cf.: Gupta in this volume). The first section of the main canal had been built during the rule of Sultan Firûz Shâh Tughluq (1351-1388). This canal was repaired and improved in 1561.

When Shâhjahân decided to build his new capital at Delhi, his architects had just completed a channel that brought water to Lahore over a distance of more than one hundred miles. He ordered them to restore the Firûz Shâh Canal and to extend it to the new city. The canal ran through the outskirts of the city, watering gardens and fields. It entered the city by the Kabuli Gate in the north-western part of the city and then split into two branches. One branch flowed down the middle of Chândnî Chawk. The other one passed through the gardens north of Chândnî Chawk and then entered the palace near the Shâh Burj. Water flowed in a marble channel through all the buildings on the eastern wall. Other channels provided water for the gardens, streets and houses inside the fort.

Besides the canal there were in and around Shâhjahânâbâd several wells, springs, step-wells and tanks, some of them dating back to earlier days. These wells had to provide the drinking water during periods when the canal ran dry, like in the late eighteenth century when the city lacked a sufficiently organized administration.

The Main Canal was reopened in 1821 and again provided potable water to the city dwellers. In 1846 a large tank was constructed between the palace and Khâṣṣ Bâzâr. Linked to the main canal, the Ellenborough Tank (popularly known as 'Lâl Ḍiggî') was to serve as a reservoir for drinking water. Roberts' Report of 1847 counts 678 wells within the city, though more than 500 were pronounced to be brackish according to other sources.

Gardens: Irrigated gardens as features of climatological control and as recreational areas were an important urban element with a long tradition in India (Crowe-Haywood 1972). For the pious Muslim the garden served as a reminder of the Quranic paradise (Brookes 1987, Moynihan 1982). The Mughal gardens were laid out according to the Persian chahâr bâgh concept, but were less rigid because of the fusion of various Indo-Islamic traditions and Hindu craftsmanship.

Most of the gardens of Shâhjahânâbâd lay in the vicinity of the city outside the city walls. The most notable of these gardens built in the 1650s were Shâhjahân's Tîs Hazârî Bâgh just outside Kabuli Gate, Raushan Ârâ Begum's garden near Lahori Gate, Nawâb Sirhindî Begum's garden in the same area, and Nawâb Akbarâbâdî Begum's Shâlîmâr Bagh six miles beyond Lahori Gate, which resembled the Shâlîmâr Bâghs in Lahore and Kashmir.

Many large courtyard residences (hawêlîs) in the city also had enclosed gardens. But the most beautiful and at almost fifty acres by far the largest garden was the Sâhibâbâd, built by Jahânârâ Begum north of Chândnî Chawk in 1650.

Thânâs / Wards / Mahallahs: Until 1857 the British more or less kept the Mughal system of administering the city. From the kotwâlî in Chândnî Chawk the kotwâl and his twelve thânâdârs policed the city, controlled the markets and bazaars and collected taxes and duties, as in former times. The only difference was that the kotwâl was now acting under the supervision of a British official.

The city was divided into 12 thânâs (wards) each under the control of a thânâdâr. Each thânâ was again subdivided into several mahallahs (neighbourhoods). The thânâdârs maintained up-to-date statistics, including tax lists and information on the population residing in their respective thânâ.

The spatial system of the city was based on an extensive hierarchical organisation which allowed a heterogeneous population to live together. The Sharia's values were accomodated by differentiating the city's space into public, semi-private and private space. The city was separated from the surrounding land by a wall and moat. Passing through the city gates marked the passage from one dominion to another. The main thoroughfares, the secondary roads and the bazaars were public space. The mahallahs were sealed, homogenous units within the city. They could only be reached by means of several gates. The alleys in the mahallah were therefore semi-private space, while the court yard houses were private space separated once again from the outside world by a gate. Mahallahs were often referred to by the name of the individual whose hawêlî dominated them, or of the vocation of the people who lived there (the Lane of the Carpenters, the Leather-workers, the Quarters of the Cloth-printers, the Candlemakers).

Streets, Bazaars and Chawks: The main axes of the city were two major boulevards connecting the fort to the city gates. The larger and more important one was Chândnî Chawk running from the Lahori Gate of the fort to the Fathepuri Masjid. From there the road sidestepped to the north before continuing to the Lahori Gate of the city. The road was thus laid out between two focal points of the city. Chândnî Chawk was forty yards wide and contained more than 1500 shops of a uniform design. Each shop occupied one room under one section of a long arcade. One branch of the main canal with trees on either side flowed through the centre of the street. Chândnî

Chawk was divided into three sections by two squares. The first section from the fort to the rectangular square was the favourite bazaar for the members of the imperial household. To the south of the square was the kotwâlî, the seat of the city's magistrate and police. The second octagonal square established a cross axis to the north, where a large serai was constructed for privileged merchants. A large hammam was also added to the complex.

The second boulevard was Faiz Bazaar running north-south from the Akbarabadi Gate of the fort (or Delhi Gate) to the Delhi Gate of the city. This bazaar had more than 800 shops of a similar design to those on Chândnî Chawk. The square (chawk) at the northern end of Faiz Bazaar was 160 yards long and about 60 yards wide and had a pool and fountains in the centre. To the west of this square Nawâb Akbarâbâdî Begum built a beautiful mosque and next to it a large serai. On the eastern side, opposite the mosque and the serai she constructed a hammam.

In addition to the two main boulevards another important bazaar was laid out between the fort and the Friday Mosque. The Khâṣṣ Bâzâr was also divided into two parts by another chawk. It was a popular bazaar crowded with healers, story-tellers, astrologers and dancing girls. Every Friday the royal procession attending Friday prayers had to pass through this bazaar.

Besides these three main bazaars several others existed all over the city. Along secondary roads specialized bazaars developed in close association with workshops or kârkhânâs. Most of the mahallahs also had local bazaars or markets serving the needs of the neighbourhood.

Fort/Palace: Just as Shâhjahânâbâd was divided into imperial (fort/palace) and ordinary space (city), the fort was similarly divided. Its axes were precisely aligned with the cardinal points of the compass. In the centre was the public audience hall (Dîwân-i Âm). West of it was the ordinary space, open to the public. Here were the bazaars, the imperial kârkhânâs, the offices and the two gateways to the city, Lahori Gate and Akbarabadi Gate. An architectural innovation in India was the covered bazaar between the Lahori Gate and the square in front of the Naqqar Khânâ. The members of the imperial household residing in the palace came here to shop and stroll shaded from the sun.

Special permission was needed to go beyond the west side of the Dîwân-i-Âm. North of it were the palace gardens which again were laid out according to the classical Persian chahâr-bâgh concept. East of it was the private palace area with the marble appartments along the river front. South of the Dîwân-i-Âm was the most private part of the fort, the harem. Throughout the fort, every detail is formal and regular, based on an extensive grid of squares. Within this plan, each building is placed according to its function. The intimate relationship between the delicate buildings in the private area of the palace contrasts strikingly with the great public vista that leads from the Lahore Gate through the covered bazaar and the Naqqâr Khânâ to the Dîwân-i Âm. In the city this central axis of the fort continued into the wide boulevard of Chândnî Chawk to the Fatahpurî Masjid and even beyond to the Îdgah outside the city wall.

Mosques: The religious infrastructure of Shâhjahânâbâd comprises hundreds of mosques and temples, shrines and khânqâhs, religious endowments, ghâṭs etc. It is an interesting feature of this mid-nineteenth century map, that information on the religious institutions of Hindus, Sikhs or Jains seem to be missing, whereas most of the mosques are shown, the prominent ones even in beautiful detail.

Shâhjahânâbâd was built around the twin foci of the palace/fort and the Jama Masjid. Erected on a hillock about 500 yards south-west of the fort and well above the surrounding city, it is one of the largest mosques on the subcontinent. The foundations of the Jama Masjid were laid in 1650. The mosque proper stands on top of a massive sandstone terrace placed in a system of radiating axes. The courtyard of the mosque is reached on three sides, east, north and south, by three broad flights of steps. The eastern gate, facing the fort/palace, was the main gate or the "royal entrance". A madrasa and a hospital were built together with the Jama Masjid and the mosque was richly endowed by Shâhjahân. Unfortunately no detailed information on the original waqf properties is available. Today the Jama Masjid is part of the "Composite Jama Masjid Wakf". This waqf comprises 79 mosques, agricultural land in the vicinity of Delhi, 27 shops, 42 residential units, 4 godowns (warehouses) and 2 garages. In 1987-88 the income of this waqf was about 62,066 Rs.

Besides the main mosque of the city there were several other important mosques built by prominent members of the royal court. All but one of these "secondary" mosques were located on the two main boulevards of the city. Second in rank to the Jama Masjid was the Fatahpurî Masjid built by Nawâb Fatahpurî Begum at the western end of Chândnî Chawk in 1650. It is built of the same red sandstone which was used for the fort and the Friday Mosque. This mosque was also richly endowed with waqf properties. Behind the mosque to the west was a large serai which was part of this waqf. The courtyard of the mosque is surrounded by a series of single and double storeyed buildings. A madrasa, shops, workshops and residential appartments occupied these buildings, which were also part of the Fatahpurî Wakf. Today the Fatahpurî Wakf comprises one madrasa, a secondary school, a public library, 6 additional mosques, 97 shops, several warehouses, and more than 300 rooms and appartments. The annual income of this waqf is about 200,000 Rs.

The mahallah or neighbourhood mosques were the third group of mosques. Of local importance, they were at the bottom of the hierarchy and lay scattered all over the city. They served the people in their immediate vicinity and were built by prominent or wealthy residents of the respective mahallah or by guilds of merchants or artisans. All of the smaller mosques received their income through religious endowments.

Hawêlîs: The members of the imperial household who lived outside the fort/palace built large mansions (hawêlîs) on the model of the imperial design of the red fort.

As a rule these city palaces accommodated not only the owner and his family, but also their numerous followers, servants and craftsmen with their workshops (kârkhânâs). The internal organisation of the space within the hawêlîs was therefore also based on the strict distinction between public, semi-private and private space.

In summarizing the afore-mentioned physical patterns and structures of 19th century Shâhjahânâbâd, a formal analysis of the city and its morphology leaves no doubt: Shâhjahânâbâd has all the elements and ingredients of a typical "Islamic city". Such an observation holds true not only for the city's formal characteristics, but also for its institutions and very obviously also for its urbanity (see: contribution by Malik in this volume).

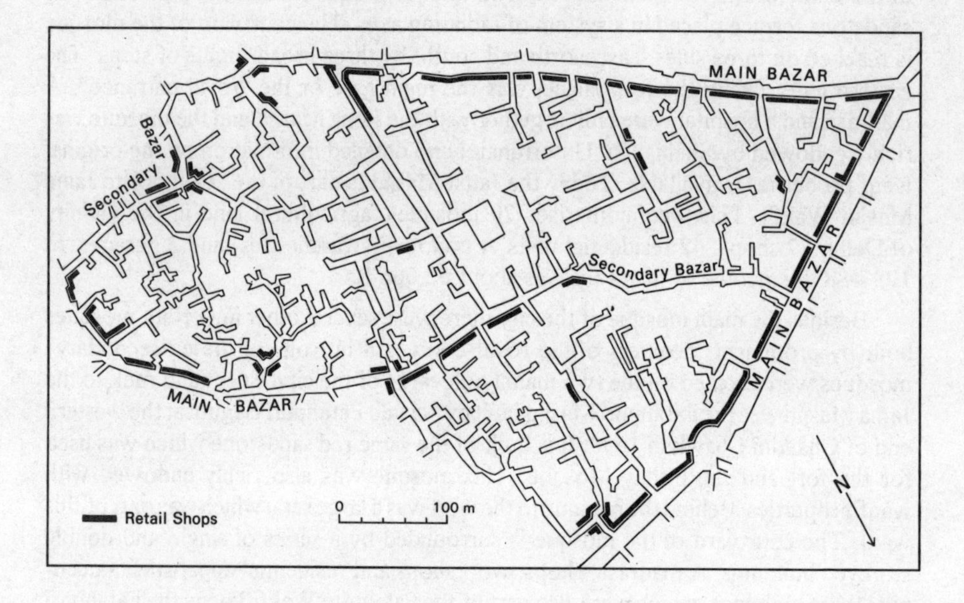

Fig. 2: A Present-Day Municipal Ward of Old Delhi with Approximate locations of Retail Shops in 1969/70 (Source: Oldenburg 1978)

Since the mid-19th century two major transformations have taken place in Shajahanabad/Old Delhi. First, the Great Mutiny of 1857 resulted in major destruction of the traditional urban fabric. Its reconstruction coincided not only with far-reaching architectural and spatial changes in the old city, but even more so in the fort and its imperial Mughal architecture. The physical transformation of Shâhjahânâbâd/Old Delhi was accentuated by the construction of Delhi's railway station, which destroyed large sections of the old city north of Chândnî Chawk. Second—and probably more important,its traditional population structure was more or less totally changed as a result of British retreat and the partition of the Indian subcontinent into an Islamic Pakistan and a predominantly Hindu-populated India in 1947. The exodus of Muslims from Shâhjahânâbâd/Old Delhi and their replacement by Hindus have, of course, greatly changed the "Islamic" character of the old city.

Nevertheless, even today a great number of physical patterns, forms and functions point to the basic and historically undoubted fact that Shâhjahânâbâd/Old Delhi was and is a city with all the attributes and prerequisites of an "Islamic city". This holds true not only for decisive aspects of its urban fabric, e.g. its street pattern (fig. 2) or its street division into mahallahs, but also for its Islamic institutions and organizations (cf. contributions by Krafft and Malik in this volume!)

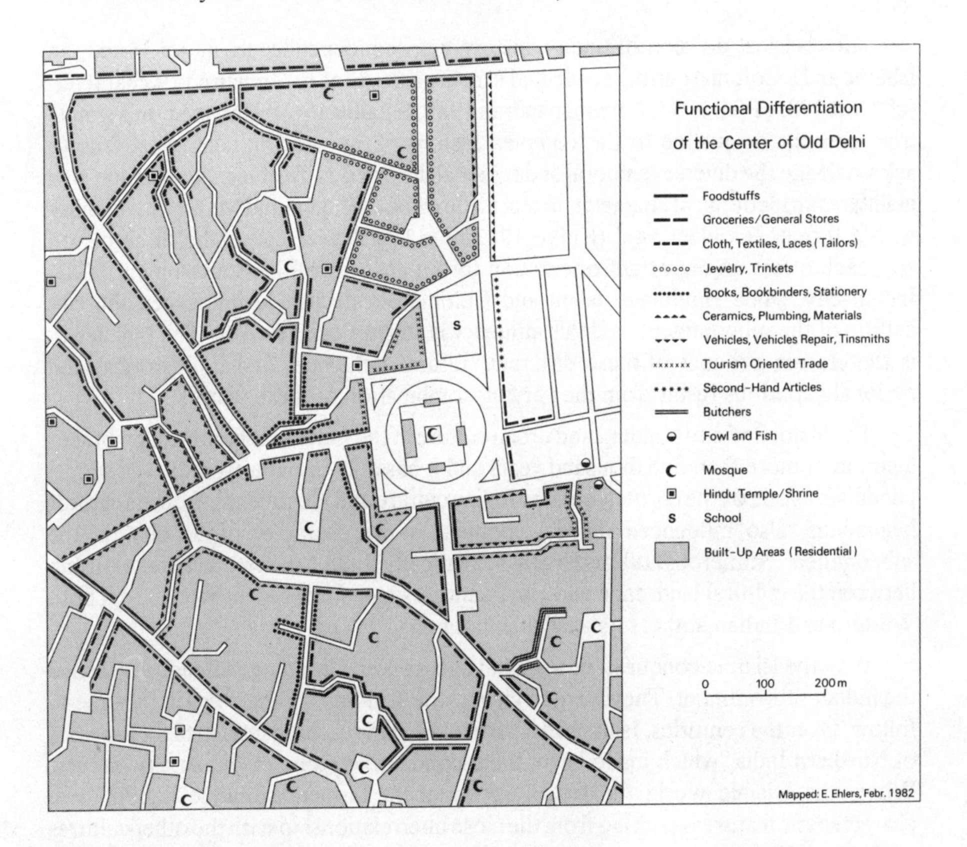

Fig. 3: Functional Differentiation of the Centre of Old Delhi in 1982

Besides these predominantly physical aspects and besides the perseverance of a great number of political and religious institutions, functional differentiations point to the fact that traditions that we consider typically "Islamic" continue to persist. Fig. 3

shows the functional differentiation of retail and wholesale around Old Delhi's Friday Mosque today. Its comparison with idealistic models of the bazaar (fig. 1) demonstrates extensive similarity: a high degree of specialisation, spatial concentration of certain branches of trade and services and—to a certain degree—even a centre-periphery differentiation within the shopping area of the old city. The fact that Dettmann's observation of the predominantly "linear" character of this bazaar type also holds true for Old Delhi, is no more than a minor regional variation of the general phenomenon.

"Islamic Cities" in India and/or the Indian Subcontinent?

The classical division of India's history before independence, into a Hindu, an Islamic and a Colonial period is reflected in many aspects of the Indian city's character. (cf. Dutt 1983; Pfeil 1935; Ramanchandran 1989; Smailes 1969) However, this simple triad can not do justice to the complex history of the subcontinent, as it fails to acknowledge the diverse temporal and regional differences. Besides, this division also maintains an ideological character, in that it allows both Hindus and Muslims to identify with a former 'Golden Age' (Kulke 1982, p. 11 f). Nevertheless the three period approach offers an important orientation for an analysis of the development of the Indian city, since Hinduism, Islam and Colonialism decisively influenced the city culture of the subcontinent, and this influence is still noticeable today. The Indian city is therefore a product of these different cultural influences and its historical and regional variations result from the varying combination of these factors.

The history of city building and urban culture in the Indian subcontinent embraces a period of more than five thousand years and it has only been partly explored so far. Hinduism as one of the cornerstones of Indian culture and identity has, from its earliest beginnings, also influenced the development of the character of the city on the subcontinent. Numerous authors have systematically studied the changing relationship between the cultural landscape and Hinduism. A synopsis of recent work, from both Western and Indian authors, is available in Singh (1987).

With the Islamic conquest, new spatial concepts and building techniques reached the Indian subcontinent. They were to change the structure of the cities in the time to follow. Over the centuries, Islam had a particularly far-reaching influence on the cities of Northern India, which made Islam the second cornerstone of Indian city culture. Within the Islamic world the Indian subcontinent formed a unique region, with characteristic features resulting from the close interrelationship with the other cultures of this area. There are, however, distinct differences even between the cities influenced by Islam, at least in their formal structure.

Traditionally, the development of Islamic architecture in India is divided into three stylistic trends: Delhi sultanates, independent regional developments and the great achievements of the Mughals (Martin 1978, p. 264 f). At its prime Mughal rule greatly contributed to and coincided with the peak of Islamic architecture and city building. The synthesis of Persian and Indian traditions and influences developed its own stylistic trend, which created impressive architectural monuments. With Fatahpurî Sikri and Shâhjahânâbâd it also made a lasting impression on city design.

The question to what extent has the Indo-Islamic city also been influenced by Hindu traditions and planning concepts in both its form and its functional construction, has not been satisfactorily answered so far, despite numerous studies on the subject. (cf. Bedge 1978; Dutt 1977) This is partly due to a lack of relevant historical sources, but an exclusive classification of the individual formal elements hardly seems possible. Therefore the arguments remain for the most part speculative.

The old city of Hyderabad offers an interesting example of this problem. With its general locational features, its oriented pattern of two crossing axial streets, its more regular street system, its central principal building, Châr Mînâr, and the central position of the emperor's residence and religious focal point, it seems to reflect a definite dependance on traditional Hindu city planning concepts. However, we are concerned here with a city that is known to have Islamic foundations or, according to Pieper (n.d.) with "a Qur'anic paradise in architectural metaphors".

Conversely, Giese (1980) suspected that a series of Islamic cities in central Asia, which showed the same regular structure as Hyderabad although they originated considerably earlier, had a line of development going back to the old Indian city structures. Without unambiguous reference sources the question posed can hardly be answered satisfactorily. However these examples demonstrate the problems which can arise if formal aspects are exclusively regarded.

Again, there is no doubt that, especially in Northern India, the morphological and organizational structure of many cities and towns was strongly influenced by their Islamic rulers and also by their Islamic population. Shâhjahânâbâd/Old Delhi, as an example of a city originating during the heyday of Mughal rule and designed as the capital of an Islamic empire, is an ideal representation of the Indo-Islamic city. Other cities, such as Agra, Ajmer, Aligarh, or Lucknow, also show corresponding features, though to a varying extent.

One fundamental element affecting the physical structure and organization of the Indo-Islamic city was the legal system of the time. Citing Tunis as an example, Hakim (1986) established the influential character of Islamic law on the city's morphology. He concludes: "...all cities in the Arab and Islamic world inhabited predominantly by Muslims share an Islamic identity which is directly due to the application of Sharia values in the process of city building" (p. 137). Other authors have also repeatedly studied the connection between city structure and Islamic law (including Spies 1921; Brunschvig 1947; Wirth 1975, Abu-Lughod 1987). Especially in Shâhjahânâbâd it is true that Islamic law lastingly defined the organization and distribution of space and therefore had a direct influence on the city structure. The division of the city into small homogeneous units (mahallas) and also the subdivision of space into public, semi-private and private space, which is to be seen at every level, is in conformity with this very law system, which eventually defined the character of the Islamic city.

Bibliography:

Abu-Lughod, J. (1987) The Islamic City: Historic Myth, Islamic Essence, and Contemporary Relevance. International Journal of Middle East Studies 19, pp. 155-176

Idem (1989) What is Islamic about a City? Some Comparative Reflections. Proceedings of the International Conference on Urbanism in Islam, Oct. 22-28, 1, pp. 194-217

Asher, Catherine B. (1992) Architecture of Mughal India. The New Cambridge History of India I:4. Cambridge: Cambridge Univ. Press

Bedge, P.V. (1978) Ancient and Medieval Town Planning in India. New Delhi: Sagar Publishers

Bonine, M.E. (1990) The Sacred Direction and City Structure: A Preliminary Analysis of the Islamic Cities of Marocco. Annual on Islamic Art and Architecture, Muqarnas, 7, pp. 50-71

Brookes, J. (1987) Gardens of Paradise. The History and Design of the Great Islamic Gardens. London: Weidenfeld and Nicholson

Brunshvig, Robert (1947) Urbanisme médiéval et droit musulman. Revue des Etudes Islamiques XV, pp 127-155

Crowe, S. - Sh. Haywood (1972) The Gardens of Mughul India. A History and Guide. London: Thames and Hudson

Dettmann, K. (1970) Zur inneren Differenzierung der islamisch-orientalischen Stadt. Ein Vergleich von Städten in der Levante und im Nordwesten des indischen Subkontinents. In: Tagungsbericht und wissenschaftl. Abhandlungen - 37. Dt. Geographentag Kiel 1969, edited by W. Meckelein and Ch. Borcherdt, pp. 488-497. Wiesbaden: F. Steiner

Idem (1970) Zur Variationsbreite der Stadt in der islamisch-orientalischen Welt. Geogr. Zeitschrift 58, pp. 95-123

Idem (1980) Städtewesen und Stadtstrukturen im Norden des Industieflandes. Mitt. Fränk. Geogr. Ges. 25/26 (1978/1979), pp. 351-393

Dutt, B.D. (1977) Town planning in Ancient India. Delhi: New Asian Publishers

Ehlers, E. (1992) The City of the Islamic Middle East. In: Ehlers, E. (ed.): Modelling the City-Cross-Cultural Perspectives, Colloquium Geographicum, 22, pp. 89-107. Bonn: F. Dümmler

Ehlers, E. - Th. Krafft (1991) The Imperial Islamic City: 19th Century Shahjahanabad/Old Delhi. In: Environmental Design. (Proceedings of the 7th International Convention of the Islamic Environmental Design Research Center in Rome, July 1991) (in print)

Fonseca, R. (1971) The Walled City of Old Delhi. Ekistics 182, pp. 72-80

Fryckenberg, R. E. (ed.) (1986) Delhi through the Ages: Essays in Urban History, Culture and Society. Delhi: Oxford University Press

Giese, E. (1980) Aufbau, Entwicklung und Genese der islamisch-orientalistischen Stadt in Sowjet-Mittelasien. Erdkunde 34, pp. 46-60

Grunebaum, G.E.v. (1955) Die islamische Stadt. Saeculum 6, pp. 138-153; Engl. Tr.: The Structure of the Muslim Town, Islam. Essays in the Nature and Growth of a Cultural Tradition, pp. 141-158. London

Gupta, Narayani (1981) Delhi between two Empires, 1803-1931: Society, Government and Urban Growth. New Delhi: Oxford University Press

Hakim, Besim S. (1986) Arabic-Islamic Cities: Building and Planning Principles. London: KPI

Inalcik, H. (1990) Istanbul: An Islamic City. Journal of Islamic Studies, 1, pp. 1-23

Koch, E.(1991) Mughal Architecture. Munich: Prestel

Krafft, Th. (1993) Shahjahanabad - Old Delhi - Zur Persistenz islamischer Strukturelemente in der nordindischen Stadt. Dissertation Bonn (in Preparation)

Kulke, H. - D. Rothermund (1982) Geschichte Indiens. Stuttgart: Kohlhammer

Martin, G. (1978) Indian Subcontinent. In: Michell, G. (ed.): Architecture of the Islamic World. London: Thames and Hudson, pp. 264-274

Moynihan, E.B. (1982) Paradise as a Garden. In Persia and Mughal India. London: Scolar Press

Nurul Hasan, S. (1991) The Morphology of a Medieval Indian City: A Case Study of Shahjahanabad. In: BANGA, Indu (ed.): The City in Indian History. New Delhi: Manohar

Oldenburg, Ph. (1978) Big City Government in India. Councilor, Administrator and Citizen in Delhi. New Delhi: Manohar

Petruccioli, A. - A. Terranova (1985) Modelli culturali nell′ impianto e nelle transformazioni di Old Delhi. In: Storia della città 31-32, pp. 123-144

Pieper, Jane (n.d.) Hyderabad: A Qur'anic Paradise in Architectural Metaphors. Environmental Design O; pp. 46-51

Ramanchandran, R. (1989) Urbanization and Urban Systems in India. Delhi: Oxford Univ. Press

Schlingloff, D. (1969) Die altindische Stadt. Eine vergleichende Untersuchung. Akademie der Wissenschaften und der Literatur Mainz, Abh. der geistes- und sozialwiss. Kl., Nr. 5. Wiesbaden: F. Steiner

Singh, R. P. B. (1987) Emergence of the Geography of Belief Systems (GBS), and a Search for Identity in India. National Geographical Journal of India 33, pp. 184-204.

Smailes, A.E. (1969) The Indian City. A Descriptive Model. Geogr. Zeitschrift 57, pp. 177-190

Spiess, O. (1927) Islamisches Nachbarschaftsrecht nach schafiitischer Lehre. Zeitschr. f. vergleichende Rechtswissenschaft XLII, pp.393-421

Troll, Chr. (ed.) (1989) Muslim Shrines in India. Delhi: Oxford University Press

Wirth, E. (1975) Die orientalische Stadt. Ein Überblick aufgrund jüngster Forschungen zur materiellen Kultur. Saeculum 26, pp. 45-94

Idem (1991) Zur Konzeption der islamischen Stadt. Die Welt des Islams, XXXI, pp. 50-92

The Indomitable City

Narayani Gupta (Delhi)

The area called Delhi has been the site of many villages and many towns over the centuries. In the angle created by the outcrop of the Aravalli Hills which is locally known as the Pahârî (hill) or the Ridge, and the river Yamuna there have been cities which became known in other parts of the world, known because they were associated with great emperors and because of their magnificent monuments and ambience of opulence. Rulers have come and gone, but the basic features of Delhi have not changed. It was the hinge between the Punjab and the Doab (the land of the Five Rivers of the Indus, and the land of the Two Rivers, the Ganga and Yamuna) and was an ideal central place for empires which extended from Afghanistan to Bengal, and from Kashmir to the Deccan. It has always been a major entrepôt of long-distance trade as well as a centre of consumption. Characterised by extreme wealth as well as poverty, it has been an open city, very cosmopolitan. The sense of urbanism in Delhi has been one which has been imposed from above, not crystallised slowly out of local communities. Lâl Kot in the 12th century, Sirî in the 13th, Ferozâbâd in the 14th and Dîn Panâh in the 16th were some of the cities in the Delhi area which were the forerunners of the city of the Emperor Shâhjahân built in the 17th century. Many settlements did not last long. The first Mughal Emperor, Babur, had remarked that "In Hindustan (north India) the populousness and the decay or total destruction of cities is almost instantaneous"[1]. Shâhjahânâbâd, however, despite invasion and change of sovereign, has not become depopulated, and there are families living there today who can trace their ancestry to the seventeenth century.

Emperor Shâhjahân (1592-1666) had a passion for building. All the Mughal rulers (who were sovereigns of north India from the 16th to the 19th century) are remembered for their creativity and love of beautiful things, ranging from wild flowers to great palaces. Shâhjahân had inherited the elegant cities of Lahore and Agra, but soon after his accession "the thought came to his mind that he should select some pleasant site on the banks of the river, distinguished by its genial climate, where he might found a splendid fort and delightful edifices.... He envisioned that streams of water should be made to flow through the proposed fort and that its terraces should overlook the river"[2]. After a long survey, he decided that Delhi was the ideal place; he chose a site

---

1 The word Mughal (Mongol) indicate the antecedents of the dynasty, but since they had become part of the Chagatai tribe of Turks, Babur's family is called Turkish and not Mongol.

2 W.E. BEGLEY and Z.A. DESAI (eds.) Shahjahannama of Inayat Khan (Delhi, 1990), p. 406.

on the right bank of the Yamuna, north of Dînpanâh and Ferozâbâd, and extending to Salîmgarh (a small island-fort which had been built in 1546). As with earlier cities in the Delhi region, Shâhjahânâbâd came to incorporate sections of older settlements. The city of Sher Shâh Sur (1540s) had included that of Feroze Shâh (1350s); the northern gate of Sher Shâh's city was some yards to the south of Shâhjahân's southern gate (the Delhi Gate). The highway leading out from Feroze Shâh's citadel to his hunting-lodge on the northern Ridge became one of the two main avenues of Shâhjahânâbâd (Faiz Bazaar). One of the mosques built by the prime minister of Feroze Shâh in 1387 (the Masjid Kalan) was skilfully incorporated into the street-scape of the new city.

At the outset Shâhjahân, like modern town-planning agencies, 'acquired' a large area around his fort, much of it agricultural land interspersed with the monumental remains of older towns. Over time, sections of these were given away as gifts or rewards, what remained as royal properties being called 'nuzûl'. Many holdings were cultivated as gardens or orchards (which were accordingly exempted from paying land-revenue) and effectively created a green belt around the built area. To keep these fields and gardens green a hydraulic strategy was necessary. Near and in Shâhjahânâbâd were wells, springs, step-wells and tanks of earlier days. But what, over time, had modified the parched dry climate of Delhi was the vast canal network which had been inspired in the 13th century by the ruler's love of verdure and of sparkling water. Engineers had diverted part of the Yamuna from a point many miles north of Delhi and coaxed it into an interlaced system of channels. These were repaired in 1561. Shâhjahân's able military commander Alî Mardân Khân, who had completed a canal to Lahore the same year as the Emperor took the decision to build his capital at Delhi, repaired the old canal and cut a new channel which ran through the chief streets, gardens and palace buildings. It was this that "gave the greatest lustre and splendour to the new city"[3]. That section of the canal which was within the city walls was called Nahar-e-Bihisht (Canal of Paradise)[4]. It was somehow fitting that Shâhjahân should have nursed back to prosperity the Delhi area which in 1398 his ancestor Timûr laid waste.

Once the area of the new capital was earmarked and the supply of water assured, the built landscape developed. The magnificent skyline of the citadel and of the Jama Masjid, outlined in red sandstone and white marble, softened by the outlines of trees, had taken shape in nine years. Delhi had earlier been a rival and a counter-magnet to Baghdad. For Shâhjahân, the rival to his city was the Isfahan of Shâh Abbâs[5]. The implicit competition was in terms of size, design and grandeur. The Emperor spurred

3 Major POLIER, Extract of Letters, 22 May 1776 Asiatic Annual Register 1800, p. 37.

4 The family name of India's first prime minister, Jawaharlal NEHRU, is derived from nahar (Canal); one of his ancestors had, as an officail at Delhi, been in charge of the canal department.

5 "One day Shâhjahân stated after looking at maps of Baghdad and Isfahan, where the bazaars were octagonal and covered, and which had appealed to his fancy, that those in the new city had not been constructed accordingly". Maathir-ul-Umara by Nawab Shah Nawaz Khan, translated by H. BEVERIDGE, Vol. II, Part I (Patna, 1979) p. 270.

his engineers and craftsmen on to greater feats; the best testimony to the superb qualitiy of their work is that today those Shâhjahâni buildings which have escaped destruction or vandalization look far stronger than many buildings of later date, and in aesthetic terms are unsurpassed. Once "the auspicious fort.... was completed, all the exalted princes and the honoured Amirs arranged to build on its right and left and along the river bank grand and imposing buildings and pleasant houses. These were constructed by the poor and rich and great men according to their limited or ample means[6]. When Francois Bernier lived in the city in the 1660s, he commented on the colonies of mud- and thatch houses in the interstices of the nobles' palaces[7]. Over the decades, these houses were strengthened into or replaced by brick houses, and the network of streets and lanes took shape. The built area was modified not by official decree but because of distress or augmented wealth.

A 14th-century poet of Delhi, Amîr Khusro, had written of the city, "If on earth there be Paradise/ It is here, it is here, it is here". Shâhjahân has this verse inscribed at the entrance of one of the halls in his palace. To his, as to all the Mughals, Paradise was not just a walled garden (Para daeza) but a beautiful city. His great-grandfather Babur had been sharply disappointed when he first saw what was to be his kingdom. The country and towns of India were extremely unattractive, he had remarked. Its people had no idea of the charms of friendly society, there were "no good horses, no good meat, no grapes or muskmelons, no good fruits, no ice or cold water, no good food or bread in the bazaars, no baths or colleges, no candles or torches, not a candlestick."[8] How Babur would have rejoiced to see the horses paraded in the maidan (open area) outside the Fort, the grapes cultivated in the gardens north of the Fort, the melons sweetening on the banks of the Yamuna as the hot summer winds ripened them, the icepits south of the city, the fragrance of excellent bread in the Palace kitchens, the public baths, the well-endowed colleges, where theological debate in the morning gave place to poetry recitals in the evenings.

Shâhjahânâbâd was pre-eminently a Mughal city, but its lifestyle was delineated largely by its inhabitants. There were to be occasions when political crises led to the city being forcibly or voluntarily emptied of people, but what was more remarkable was the immigration, by individuals and communities, over the following centuries. An old saying that Delhi has many gates of entry but none for departure has always been valid. It was very cosmopolitan, with a mix of people from all parts of the Empire and beyond.[9] When state power was strong, houses used to be resumed by the government when the incumbent died, since they were not meant to be conferred in perpetuity.

---

6 Ibid., p. 271.

7 F. BERNIER, Travels in the Mogul Empire, A.D. 1656-1668 (Delhi, 1968), p. 241.

8 H.M. ELLIOT and J. DOWSON, The History of India as told by its own Historian, Vol. IV (Allahabad, n.d.), p. 222.

9 "There was no ward in which there was not the house of some Iranian Officer", Maathir, Vol. II, Part I, p. 177.

Many large hawêlîs (mansions) were built by rajas and nawabs who were in alliance with or in a state of subservience to the sovereign; by officials and bankers, and by members of the royal family. Places of worship, charitable institutions, shopping precincts and public baths were built by individuals[10]; the best known endowment (waqf) of this kind was the spacious caravanserai and gardens commissioned by the daughter of Shâh Jahân, Princess Jahânârâ, in Chândnî Chawk (= Moonlight Place). Till the mid-19th century there was a harmonious mix of open space and built up area in Shâhjahânâbâd. Shops and homes shared buildings fronting on to the streets. There was a hierarchy of markets, ranging from all-purpose neighbourhood ones to specialised clusters, like the Darîbâ Kalân (the Great Street of the Jewellers) distinguished from Darîbâ Khurd (the small Street of the Jewellers) and Kinârî Bazaar (Tinsel Street). Bishop Wilson in 1836 wrote that the distant view of the city, with its dreaming spires resembled "that of Oxford from the Banbury Road"; the simile vanished when he entered, and saw "the wide streets, the ample bazaars, the shops with every kind of elegant wares, the prodigious elephants used for all purposes, the numerous native carriages drawn by noble oxen, the children bedizened with finery, the vast elevation of the mosque, fountains and caravanserais for travellers, the canals full of running water raised in the middle"[11].

The reason why a large number of people could live together in this compact area, and accomodate newcomers without social tension being generated was that urban society was a highly regulated one, and everyone knew the rules. It was a hierarchy of Chinese boxes, ranging from the city wall to the curtained private quarters of the house. The city was formally entered through one of the gateways ("the most magnificent which the world can boast")[12], of which there had been eight in 1657[13] and fourteen, together with another fourteen khirkîs (wicket-gates) by the nineteenth century[14]. The wall had been initially made of mud, but was later reinforced by stone. The writ of the city Kotwâl (magistrate) was circumscribed by the wall. The city was divided into 12 thânâs (wards) each in the charge of a thânâdâr; each thânâ was a honeycomb of mahallahs (neighbourhoods) under the responsibility of mahallahdars. It was in the mahallah that the visitor came up against his second formal entry-point, since each mahallah was sealed off from its neighbours and could be entered only by one gate[15]. The names of newcomers to the city were maintained by the thânâdârs, who

10 "Mohammad Yâr Khân was the owner of many houses and shops in Delhi and exaggerated accounts were current regarding the high rents he used to realize for them", ibid., p. 218.

11 Society for the Propagation of the Gospel, Historical Sketches, Missionary Series, No. 1, 'Delhi', London 1891, p. 4.

12 'Mofussil Stations No. XI, Delhi, Asiatic Journal, May/August 1834, p. 2. The phrase is not Urdu hyperbole, but English understatement!

13 Begley and Desai, op.cit., p. 537

14 J.A. PAGE (ed.) List of Mohammedan and Hindu Monuments, Delhi Province Vol. I, Shâhjahânâbâd Zail (London 1913), p. 35

15 This was known as the Kûchâ-bandî system (Kûchâ = lane, bandî = closure)

also had up-to-date details of vital statistics and tax collections; cleanliness was maintained by chowkîdârs in each mahallah paid by the residents. Mahallahs were known by the name of the individual whose hawêlî dominated it, or of the vocation of the people who lived there (the Lane of the Carpenters, the Leatherworkers, the Quarters of the Cloth-printers). Hawêlîs were often minitowns, accommodating not only the family of the owners, but also their retainers, and craftsmen and artisans with their kârkhânâs (workshops). The hawêlî was entered by crossing the deorhî (threshold): indoors, space was divided into public and private, the private in turn into the zenânâ (women's quarters) and mardana (men's rooms). Formality increased as one proceeded from street to neighbourhood to house. The same strict distinction could be found in the palace-complex.

The citadel was open to the riverfront, but set at a height above the level of the water; it was separated from the city by a purdah (curtain) of red sandstone. For two centuries it was a Vatican within the Rome of Shâhjahânâbâd. The Emperor who supervised its design was able to enjoy its ambience for only ten years before his throne was usurped by his son Aurangzeb, and Shâhjahân himself was kept confined in the fort of Agra. The citadel remained as Shâhjahân had left it, except that Aurangzeb constructed an outwork at the western gateway, so that the Hall of Public Audience was not in the direct view of the Chândnî Chawk avenue; this was intended to give some respite to the people who happened to be in the avenue, since custom demanded that anyone in the view of the Emperor had to bow. An exquisite marble mosque, the Motî Masjid (Pearl Mosque) in the fort was also his contribution. Within the citadel lived the Emperor's family and dependants (one estimate puts their strength at 10.000)[16] and staff. The southern section of the citadel was the private area, the central and northern the public halls and offices. The buildings were embedded in gardens and separated by channels of water, creating a congenial microclimate "L'interior est orné des édifices, des beaux appartments, de lieux des promenades délicieux et des jardins entièrement agréables"[17].

The Palace faced east as well as west, east to the Doab and west towards Afghanistan, away from the city as well as towards it. The visitor entering the city from the west, through the Lahori Gate, entered the palace through its corresponding Lahori Gate. The palace could also be entered by the river-gate to the southeast[18]. The city slopes down to the river-bank as can be experienced in the areas to the north and south of the citadel. In the case of the palace itself, the contrast is more dramatic. From the riverfront windows of the palace the bank is far below, and the architecture of the complex becomes intelligible if one remembers that below the building is a layer

16 A. PETRUCCIOLI, 'Delhi the old city', The India Magazine, No. 7, January 1987, p. 10

17 Notes by Legendre for writing the History of India, F. 346, Nouvelles Acquisitions, Bibliothèque Nationale, France

18 (In 1669) Husain Pasha Shah of Turkey "received from the Emperor a lofty mansion on the bank of the Yamuna...and a boat so that he might come by river to the court" Maathir, Vol. I, p. 699

of tahkhânâs (basement rooms) which were cool and dark in summer. During the months of the monsoon, the water of the river would have lapped the walls of the Fort. In the arid summer months it would retreat, leaving a marshy tract. It is here that the people of the city would gather in the mornings when the Emperor was in residence, to catch a glimpse of him through a trellis window. This practice of jharokhâ-darshan (window-viewing) was an old Indian ritual, symbolic of the fact that the rulers were at a higher level than their subjects, but also accessible to them. The view of the citadel from the riverfront was spectacular. In 1833, an English visitor catching his first glimpse of the Mughal palace from the river was to rhapsodize "With the rising sun glittering on its numerous marble minarets and gilded domes it was a most gorgeous spectacle"[19].

Accounts of the palace and of the city at their most opulent must have been even after making allowance for verbal embellishment. The city seemed to glitter at the centre of the great Mughal empire as the Koh-i-noor diamond given to Bâbur glittered at the centre of the jewel-encrusted Peacock Throne that stood in Shâhjahân's Hall of Private Audience.

A near-contemporary account vividly describes the Palace. "The lofty fort, which is octagonal according to the Baghdad style, is 1.000 royal yards long, and 300 yards broad. Its walls are built of the red stone of Fathpur. Its height including the battlements, from the foot of the wall is 12 1/2 yards. Its area is six lac yards, which is double that of the great fort of Akbarabad (Agra) and its perimeter is one thousand six hundred and fifty yards. It has twenty one bastions, seven circular and fourteen octagonal; four gates and two windows. Round it is a moat twenty yards wide and ten yards deep; this is supplied with water from a canal connected on two sides with the river Yamuna—except on the east side where the wall of the fort abuts directly on to the river—it was built at a cost of twenty one lacs of rupees. The royal mansions, consisting of the Shâh Mahal with a silver roof, Imtiyaz Mahal with the bedroom known as the Burj-Tâlâ (The Golden Chamber), and the private and public Daulat Khânâ (Palace), and the Hayât Bakhsh garden cost twenty eight lacs of rupees. The palaces of the Begum Sahiba and other chaste inmates of the Harem cost seven lacs, and other buildings, such as the bazar and the guard-houses inside the mighty fort, which were designed to serve for the royal manufactories, were completed at a cost of four lacs.

"On 24th Rabe I, 1058 A.H. (8th April, 1648 A.D.) in the 21st year of the reign, the day which had been selected by the astrologers for royal entry, orders were issued for arranging the paraphernalia of a royal feast and a convivial entertainment. In all the royal apartments were spread beautiful carpets, which had been prepared in Kashmir and Lahore out of selected wool with great skill and taste, while on the doors of the courtyards and porticoes were hung curtains embroidered, worked in gold, and velvet

19 Diary of Captain de Wend, 11 September 1833, Centre for South Asian Studies, Cambridge, n.p.

brocades prepared by the skilled workers of Gujarat. In every apartment were placed jewelled, gold enamelled, and plainly worked thrones, and after arranging high seats and cushions with covers of brilliant pearls, gold embroidered cloths were spread over the thrones. The three sides of the great portico of the private and public palaces were embellished with a silver enclosure, and opposite the Jharoka was a golden enclosure, while golden stars with golden chains were hung in all alcoves, and these made the palace resemble the heavens. In the middle of that portico was placed a square throne surrounded by a golden enclosure; the heavenly jewelled throne was left exposed to the sky whence the splendour of the world-illuminating sun radiated. In front of the throne was erected a canopy embroidered with gold and pearl strings, and raised on jewelled poles; and on the two sides of the throne were placed two parasols (chatr) decorated with pearl strings, while on the other two sides of the throne octagonal frames were erected. Behind the throne were placed jewelled and golden tables on which was displayed the Qur Khânâ—which consisted of the jewelled swords with worked scabbards, quivers and gem-bedecked arms, and jewelled spears for the making of which full use had been made of all the resources of the sea and the mines. The roof, the pillars, the doors and walls of this heavenly portico, and the porticoes all round the private and public palaces were covered up and decorated with embroidered canopies, golden curtains from Europe and China, gold and silver embroidered velvets from Gujarat and gold and silver thread screens. In front of the great central portico was erected an awning of gold embroidered velvet, and in f ront of the lateral porticoes canopies of embroidered velvet with silvered poles, and having spread coloured carpets on the floor of this canopy a silver enclosure was erected round it. This great canopy, which in its height and extent resembled the heavens, was, according to royal orders, woven in the imperial factory at Ahmadabad, and took a long time to complete at a cost of nearly one lac of rupees. Its length was 70 royal cubits and width 45 cubits. It was erected on four silver poles, each of which was two yards and a quarter in circumference and 22 yards high. It covered an area of 32.000 (square) yards and 10.000 people could be accommodated under it. It took trained farashes and 3.000 additional men working hard for a month to erect it and it was generally known as Dalbadal. In short, such a canopy—which resembled the heavens—had never been erected before and such a building—which was a counterpart of the heavens—had never been decorated so elaborately. From the date of the auspicious entry of the Emperor into this heavenly building there was a continuous, grand feast lasting ten days. Every day a hundred people were exalted with the grants of suitable Khilats, many were gratified by an increase in their ranks and the granting of titles, while others received grants in cash, horses and elephants."

After the death of Emperor Aurangzeb (in 1707) the rulers were less able and the treasury less replete than earlier. Advantage was taken of this by other rulers. The Mughal Empire suffered invasion for the first time in 1739, from the forces of the Persian Emperor Nâdir Shâh. Like Timûr in 1398, he attacked not to conquer but only to loot. He carried away many camelloads of wealth from Delhi, the Peacock Throne crowning it all. He also ordered a massacre, which filled Chândnî Chawk with corpses and made his name in north India comparable with that of Wallenstein in central Europe in the 17th century. Once Nâdir Shâh showed that Delhi was vulnerable, others were not slow to join in. Delhi was ravaged successively by the rebellious governor of Awadh (east of Delhi), by the Afghans from the northwest, by the chiefs of Rohilkhand (to the northeast), the Marathas (from the southwest) and the Jats (from the region immediately south). Emperor Shâh Âlam II reigned as long as Emperor Aurangzeb had done (1759-1806), but for some years he deemed it wise to escape from Delhi and live in Allahabad. The Marathas were defeated by the troops of the British East India Company in 1803 at the battle of Patparganj (across the river from the palace) but the Mughals found that they had only exchanged one master for another. Delhi, with Agra, formed part of the British booty[20].

It is only by appreciating the very fraught history of the city in the 18th century that we can reconcile the highly divergent pen-portraits that have been drawn of it. Behind the accounts of wealth and beauty, or of dilapidation and demoralization, we can see the silhouettes of a harrassed citizenry living in dread of the unknown, constantly renovating and rebuilding, hiding their treasures, suffering the anguish of seeing lovely facades vandalised, carefully-collected libraries ransacked, the life-giving canal choked up, trade paralysed. Cruelty, greed and injustice are as much embedded in the city's collective memory as are camaraderie, joie de vivre, courtoisie and a sense of the beautiful and appropriate. The essence of the Delhiwala's spirit was nostalgia, a sense of fatalism and also a will to survive. After each crisis, there were many who fled the city, and began life anew at the courts of Lucknow, Jaipur, Bhopal or Hyderabad. But some, like the poet Mîr, felt homesick and returned. And once the natural flow of trade was resumed, and prosperity returned to the markets of Delhi, the students, the merchants and the artisans also returned, and life went on again...

When Indians wrote about Delhi, they did so in terms very different from those used by Europeans. For the latter, the yardstick for comparison was Versailles, Paris, London or Moscow. For Indians the artefacts mattered less than the urban way of life and the individuals whose learning, art or wit commanded respect or affection. The court patronised artists and poets, and many of the emperors were multilingual, with a taste for reading and even for versifying. Chândnî Chawk was, like the Italian Piazzas, less a place than a way of life. Its qahwâ-khânâs (coffee-houses), its shops, the canal, made it the natural centre of the city. "The nobles, irrespective of their status, are unable to suppress their desire of taking a stroll here...The paths of the chowk are broad as a wide forehead and bountiful like the blessings of God"[21]. In Muraqqa-e-Delhi, a Persian work written just after Nâdir Shâh's devastation, what comes across most vividly is the people's sense of living for the moment, their eagerness to find

20 P. SPEAR, Twilight of the Mughuls (Cambridge 1951)

21 Dargah Quli Khan, Muraqqa-e-Delhi, translated by C. SHEKHAR and S.M. CHENOY (Delhi 1989), p. 25

pleasure by patronising dancers, muscians and sufi poets, and fulfilment by visiting shrines. His descriptions of women artists and their followers would be the equivalent of the diary of a visitor to contemporary Paris, writing of the salons of Mme Maintenon or Mme Pompadour. "Music was the most popular form of entertainment...It was patronised at the Imperial Court, in the establishments of the nobility, in the Khânqahs (shrines) of the living sufis, in the houses of the musicians and on the streets. Its votaries included Hindus and Muslims, sufis, saints, sultans, rich and poor. Men and women from different backgrounds and social situations."[22] Poetry was also a great leveller. In times of plenty, they wrote elegant verse; in times of sorrow, they wrote even more memorably, in the genre called shahr ashob (lament for the city) which described, sometimes with anguish and sometimes with a wry detachment, the travails of a city overtaken by political misfortune.

> These who once rode elephants now go barefoot
> And for want of a pair of shoes wander about disconsolately.
> Those who yearned for parched gram once
> Are today owners of property, palace and elephants as
> marks of rank.[23]

Hâtim's sarcasm finds no echo in the sorrow of Mîr, whose poem is also the product of the same situation.

> Delhi, which was counted among the great cities of the world
> Where lived the elite,
> Circumstances have looted and destroyed it.
> It is to that desolate city that I belong.[24]

The sense of order and of hierarchy in the city was upset after political crises. Major Polier, writing in 1776, admitted candidly that a few decades earlier "the furthest I could have pretended to go would have been about the gate" of the very hawêlî (the house of the minister Qamaruddîn) which he was living in[25]. In these situations it was misleading to judge by appearances, because many owners deliberately kept the facades of their houses in a shabby condition so as to deflect the attention of marauders. Invariably the great scourges were followed by the lesser; the Gûjars and Mewâtîs of some of the villages near Delhi used to take advantage of the distress in the city to make their own quick forays through the gateways which in times of peace were well-guarded and which kept them at bay.

---

22 Ibid., p. xxxii

23 Ibid., p. xxxix

24 S. Nurul HASAN, 'The Morphology of a Medieval Indian City, a Case-Study of Shâhjahânâbâd' in I. BANGA (ed.) The City in Indian History (Delhi, 1991), p. 96

25 Major POLIER, loc, cit., p. 29

26 N. GUPTA, 'Delhi and its Hinterland' in R.E. FRYKENBERG, Delhi Through the Ages (Delhi, 1986), pp. 250-269

Shâhjahânâbâd had an organic relationship and a complementarity with its shrines and monuments, as well as with its groves and gardens[26]. The visitors who filled the serais (inns) within the city and beyond the walls were drawn not just by the court and the salons. Shâhjahânâbâd was not a cultcentre, though the major mosques—the Jama, Fatahpurî, Akbarâbâdî and others formed part of all tourist circuits. The city was within easy reach of some historic shrines—those of Nizâmuddîn Auliya, Roshan Chirâgh Delhi, Bakhtiyâr Kâkî and Hazrat Bâqibillah; and the temples of Kâlka-Devî and Jogmâyâ. Both shrines and temples had special feast-days, which constituted a familiar calendar. The royal family regularly visited the mausoleum of Humâyûn, near the Nizâmuddîn shrine, and the palaces of Mehrauli, near the shrine of Bakhtiyâr Kâkî. Clusters of markets expanded outside the western wall of the city, to avoid the time-consuming payment of town-duties. If the repeated crises had not driven the inhabitants firmly into the shelter of the city walls, Shâhjahânâbâd would probably have grown as an extended open city as Ferozâbâd had in the 14th century, stretching from the southern Ridge to the river. Between the hamlet of Mughalpurâ and the city was the Sabzî Mandî (vegetable market). Tanners and dyers had their colonies at the base of the Ridge to the south-west. In the 18th century a spacious college was built adjacent to the tomb of Ghâzîuddîn, a nobleman of Awadh, outside the Ajmeri Gate of the city.

From 1803 to 1857 Delhi was part of the territories under the East India Company on the same terms as governed the Company's relations with states which had accepted political subservience and military dependence. The Company struck coins in the name of the Mughal Emperor, but referred to him pointedly as the 'King of Delhi'. It became increasingly evident that he was not even king of Delhi, but only within the walls of his palace. The city was virtually ruled by the British Resident, living in the palace of Dârâ Shikoh, inside Kashmiri Gate. The Mughal royal family resigned themselves to the position of 'Pensioners"; they occasionally murmured that their allowances were very meagre, but on the whole they were content to fill out their days flying kites, composing verses and keeping up a sad mockery of the ceremonies which earlier had symbolized such great power. The Resident played his part, scrupulously adhering to ritual (always riding behind the Emperor and not before him) while creating an alernative centre of power, where Mughal princes, Maratha chieftains and neighbouring rajas shared a common style of leisure pursuits with the British officials.[27]

This 'Pax Britannica' (as the officals described the years of armed peace till 1857) saw Delhi acquiring a vigorous new lease of life, its location giving it a major role as an entrepôt for subcontinental trade, and its traditions helping to create an atmosphere congenial to scholarship and artistic creativity. The old-established banking and merchant families were augmented by others from Punjab and Rajasthan. Delhi became a busy distributing centre for Indian wheat and British textiles. The steady, if undiscriminating, patronage of British tourists and of buyers for European markets

27 SPEAR, op. cit. chapter 7

meant a boom for local crafts and skills, particularly in ivoryware and in miniatures which were very convincing lookalikes of Mughal master-artists. Local shopkeepers benefited from the stationing of British soldiers and civilians north of Kashmiri Gate, in the shadow of the Ridge, and a market developed along the stretch of road which was nicknamed 'Khyber Pass".

Contemporary accounts indicate that students from many of the smaller towns in north India went to Delhi to study general courses, or skills like calligraphy and medicine from the well-known savants of the city[28]. The madrasah (school) attached to the mausoleum of Ghâzîuddîn (founded in 1792 and located in a large campus beyond Ajmeri Gate) overtook all the other educational institutions, and was chosen as the local beneficiary for a share in the small sum of money the East India Company set aside for encouraging education, in 1824. It is significant that while Bombay, Madras and Calcutta accorded an enthusiastic reception to European learning and to the English language the people of Delhi were adamant that they wanted to learn European science and philosophy, but in Urdu rather than English. The principals of Delhi College, who included a German and a Frenchman, acceded to this, and English classes formed an optional course of studies. In the three decades before the Rising of 1857 the staff and students of the College formed a dedicated community who translated many European works into Urdu, and showed a great enthusiasm for science. Its alumni included mathematicians, novelists, poets, educationists and civil servants[29]. Their efforts at the time were helped by the fact that in the 1840s a printing press was set up in Delhi which published newspapers and books. One of the greatest Urdu and Persian poets, <u>Gh</u>âlib, was writing at this time, and also seeking ways of making enough of an income to live in reasonable security. "Delhi was an Indian Weimar", Percival Spear was to write, "with <u>Gh</u>âlib for its Goethe".[30]

The camaraderie between Europeans and Indians, both social and academic, was expressed in the formation of the Delhi Archaeological Society in 1847.[31] Its members discussed various aspects of Delhi's historic architecture, discussions which spurred one of the members, a young man called Syed Ahmad <u>Kh</u>ân, to write a comprehensive account of Delhi's monuments, Âsâr-al-Sanâdîd[32]. When in 1870 the newly-estab-

28 <u>Sh</u>âh Abdul Qâdir, Syed Mohammed Amîr Rizvî, Muftî Sadruddîn Âzurdâ, <u>Sh</u>âh Abdul <u>Gh</u>anî and <u>Sh</u>âh Mohammad Ishâq were well-known scholars of Delhi. I am indebted to Mr. Rizvi of the Anglo-Arabic School, Delhi, for this information.

29 N. GUPTA, Delhi Between Two Empires 1803-1931 (Delhi, 1981). Chapter One.

30 P. SPEAR, op.cit., p. 73

31 N. GUPTA, op.cit., p. 8

32 The first edition of Âsâr-al-Sanâdîd was published in 1847; a considerably abridged and modified edition was printed in 1854. Syed Ahmad <u>Kh</u>ân later achieved distinction (and was knighted) for founding the Aligarh University. In 1829 was published Sangin Beg's Sair-ul-Marazil, a Persian travelogue which described Delhi's historic monuments and also provided what can be called a street directory for <u>Sh</u>âhjahânâbâd. This was translated into Urdu only in 1980, by Naim Ahmed (Aligarh 1980). Surprisingly enough, Syed Ahmed <u>Kh</u>ân makes no reference to this book.

lished Archaeological Survey of India prepared a report on Delhi, this was based largely on Syed Ahmed Khân's work. His book was interesting on two counts—first, its content: it described not only the buildings but also the well-known personalities of Delhi, making it not an archaeological treatise but an account of a living city and its past. Secondly, the book was remarkable in being dedicated not to the Mughal Emperor, as convention would have dictated, but to Thomas Metcalfe. This person, who spent the best years of his life in Delhi (1813 to 1853) was the brother of Charles Metcalfe, Resident at the Delhi Court. Thomas succeeded him in this post, and his son, in turn, was to be Chief Magistrate at Delhi. In other words, a Metcalfe dynasty was ruling Delhi along with the Mughal: Thomas Metcalfe was to compile his notes on Delhi in 1844 as "Reminiscences of Imperial Delhie", at the same time as Syed Ahmad Khân was completing his book[33]. The latter was illustrated with competent line-drawings, but Metcalfe's manuscript, dedicated to his daughters, was a more lavish affair, with beautiful paintings by some of Delhi's best artists. Together the work of these two Delhiwalas constitutes a discovery of Delhi which was also indicated in the interest in Delhi's older monuments by the engineers of the garrison stationed there.[34]

The city was also changing. The Metcalfes, the city Surgeon Dr. Ludlow, and other Englishmen were quite happy to live in the city, but some of them chose to build substantial homes for themselves in the area between the Kashmiri Gate and the cantonment on the Ridge. Of these Metcalfe House was the most opulent, a stately home set in spacious gardens sloping down to the river, a fascinating mixture of European and Indian architecture and lifestyle with large rooms, surrounded by long verandahs and provided with basement rooms; with formal English dining-table and equally formal Indian hookah. Metcalfe, like most well-off people of Delhi, liked to spend some weeks at the little town of Mehrauli, near to Qutab Minar; unlike them, he did not build a house or palace for his own use, but converted an old Muslim mausoleum into a weekend resort. He also used to take friends to Shalimar Gardens, a Mughal garden north of the city, where nearly two centuries earlier Emperor Aurangzeb had been crowned. Charles Trevelyan, an enterprising young civilian, did not build a house himself but bought land west of Shâhjahânâbâd and laid out a 'suburb' with plots neatly squared off for shops and houses. His Trevelyanganj survived for a long time before it was merged into the more crowded extensions that developed there. Near this was another area laid out by Dîwân Kishan Lal, a wealthy Indian, and named after him as Kishanganj. Within Shâhjahânâbâd, one of the major modifications was the addition of a large house and garden on the Chândnî Chawk avenue near the Palace, by the Indian widow of Walter Reinhardt 'le Sombre", popularly known as Begum Samru. Another large estate was carved out within the wall near Kashmiri Gate by Colonel Skinner, a gentleman of mixed birth, who also built near

33 M.M. KAYE (Ed.), The Golden Calm - Reminiscences of Emily, Lady BAYLEY and by her father Sir Thomas METCALFE (Exeter 1980)

34 J. KEAY, India Discovered, London 1981, pp. 177-187.

and facing the entrance of his house the beautiful St. James' Church where, among many others, the daughter of Thomas Metcalfe, Emily, was married. European architectural details, colonnades and tympana and grills, in freestanding houses, began to appear in the Indian areas too. Shâhjahânâbâd was becoming an Indo-Anglian town.

Bahâdur Shâh, better known by his poetic pen-name 'Zafar' (Victory) succeeded to the Mughal throne in 1837. Viceroy Dalhousie decided that he should be the last king, and that his successor should be asked to retire to Mehrauli. But the Mughal line ended in a way different from what had been planned. On a very hot day in the summer of 1857 a contingent of rebellious soldiers from the cantonment-town of Meerut, across the Yamuna, reached the palace of Delhi and urged the startled old Emperor to lead their revolt to remove the British[35]. His reluctance was brushed aside, and a council of soldiers was installed in the palace to conduct operations in his name. A large number of mutinying regiments converged on the city. They did not have the advantage of gunpowder, because very soon after the revolt began, two British soldiers in Delhi had performed a courageous act of self-sacrifice by blowing up the powder-magazine, located near Kashmiri Gate, and themselves with it. The city was in a state of siege, with the British troops encamped on the northern Ridge. It was four months before the city was captured, and the long wait and the heat had left the British with little inclination for mercy. When their soldiers stormed the Kashmiri Gate, they opened the floodgates of a reign of cruelty and terror which the dazed inhabitats compared to that of Nâdir Shâh. The happy camaraderie of the foregoing years had evaporated like a dream.

The officals' obsession with security, with averting any repetition of 1857, affected not only the ethos but also the morphology of Shâhjahânâbâd. The Palace-complex (hereafter called the Red Fort) was emptied of its inhabitants. The aged Emperor, after a long drawn-out trial, was exiled to Burma. The thousands of dependents were turned out, and they sought refuge in the back alleys of the city, and near the Qutab and Humâyûn's mausoleum. The army moved into the Fort, bulldozed most of the palaces and erected tall dreary barracks. They cleared an arch 500' wide in front of the Fort and another around the city wall, to provide a shooting-range, on the pattern of the Maidan around Fort William in Calcutta. Many houses, shops, public buildings and the beautiful Akbarâbâdî Mosque were destroyed in the process, and the organic link between the palace and the city was broken. The major mosques were confiscated, and desecrated by being used for secular purposes, one of them as a bakery. All the royal properties in the city and beyond were appropriated by the Government[36].

---

35 There is a copious literature on the Revolt of 1857, but to date there is no good scholarly work on Delhi in 1857.

36 N. GUPTA, op.cit., pp. 31-32

Delhi was made to forget that it was a Mughal City. It was made part of the Punjab, the capital of which was Lahore. The office of Kotwâl was replaced by a Municipality, with British officals and 'loyal' Indians, and by a police-force for which the local people had to pay. The gates of the mahallahs were demolished, to make passage for soldiers or policemen easier. Premises for the Delhi College (where Jahânârâ Serai had been located on Chândnî Chawk) was built in the 1860s, but instead of housing the college, it became the municipal office, with the totem of a clocktower in front. The railway-line from Calcutta to the northwest frontier was constructed through Delhi, with the station within the city instead of being to the north, or east of the river, as originally planned. Less than twenty years aftert the Rising of 1857, Delhi was a changed city; entered from the railway-station rather than by boat from the northern Kashmiri Gate, the Italiante arches of the railway station and the wide avenue of Queens Road made it difficult to visualize the crowded mahallahs that they had displaced. The term 'Queens Gardens' used for what was left of Jahânârâ's gardens made one think of Victoria and not the Mughals. Much of the western wall had been demolished, and the city overflowed into the jerrybuilt Sadar Bazaar, a whole sale shopping area which grew in response to the location of the railway-station. The Delhi College was reduced to a school, when the government chose to transfer its grant-in-aid to the missionary College of St. Stephen. This was located initially on Chândnî Chawk and then near St. James' Church. Though the churches that served the needs of the Europeans were all within the city, the houses of the civilians and the missionaries were all to the north, in the sprawling area called the 'Civil Lines' (the conventional term used for the British enclaves in all Indian towns). The bankers and merchants who had played safe during the Rising took advantage of the low prices to invest in real estate. Urban land acquired a market-value which it had never had under the Mughals. The nouveaux-riches as well as the nouveaux-pauvres were a feature of Delhi as a result of the crisis of 1857, though social status did not correspond to income levels; even impecunious aristocrats were treated with deference, and much of the new wealth was translated into donations to schools and hospitals rather than displayed in conspicuous consumption. Though the Court had been ended, the bond of the language, Urdu, united the people of Delhi, transcending religious and class difference.

An often-repeated remark about the Delhiwala sees him as happily indifferent to anything beyond his immediate area. It certainly took many years for the leaders of the Indian Nationalist Movement to find any following in Delhi. The rising political temperature in Bengal persuaded the British to shift their capital from Calcutta to the more tranquil Delhi in 1912. On three occasions, in 1877, 1903 and 1911-12, the supremacy of the British sovereign in relation to the Indian princes, had been proclaimed at elaborate 'Durbars' an Indian word appropriated by the British to imply both imperial space and imperial ceremony.[37] The Durbars had been held in a large area

---

37 B. COHN, 'Representing Authority in Victorian India', An Anthropologist Among the Historians (Delhi 1987), pp. 632-682

northeast of Shâhjahânâbâd, with supplementary processions through the city, with all the trappings of a Mughal ceremonial. Since the British saw themselves as having Mughal mantle, Delhi seemed a natural choice for a new capital. The team of architects who finally designed New Delhi were as reluctant to recycle Shâhjahânâbâd as Emperor Shâhjahân had been to recycle the city of his grandfather and father at Agra. They wished to build a completely new city, adjacent to Shâhjahânâbâd but without any organic link with it; as it was designed, it faced the river and was backed by the Ridge, just as Shâhjahânâbâd was.[38] During the years that this new city was being built (1914-1931) the bureaucratic apparatus of the Government of India spent five months every year in 'temporary' quarters on the northern Ridge. Shâhjahânâbâd had a greatly enlarged population billeted on it, and had politics thrust upon it in the form of demagogues, political conferences and protest-meetings. Chândnî Chawk became a venue for these, and Shâhjahânâbâd—henceforward called 'Old Delhi'—constituted a pulsating central zone between two tranquil belts, the Temporary Capital and New Delhi. The people of the older city increasingly had a sense of being neglected or even drained in order to nurture the more privileged townships.

In the last half-century the morphology of Greater Delhi has become well-defined, and the successive stages of historical development are evident. The area of Shâhjahânâbâd and the adjacent Sadar Bazaar constitute the most densely populated central area; the New Delhi of Lutyens is the most sparsely inhabited. Both are surrounded by settlements which have developed in the short span of forty years. After India became independent in 1947, Delhi, like many other towns of north India and the new country of Pakistan, saw a population turnover; many thousands of Muslims migrated to Pakistan and many Sikhs and Hindus came to Delhi.[39] Many houses in Shâhjahânâbâd became emptied of occupants, while around the twin cities there grew the 'refugee colonies' most of them on the land that had been royal (nuzûl) property under the Mughals and had later been under Delhi Municipality. Shâhjahânâbâd remained the major wholesale and retail trade centre it had always been, but the quality of life became degraded by the invasion of motor cars and buses, by the proliferation of numerous industrial activities in sections of the buildings which had once been gracious hawêlîs or spacious katrâs (shopping precincts)[40]. Today one can walk

38 R.G. IRVING, Indian Summer-Lutyens, Baker and Imperial Delhi (Yale 1981)

39 V.N. DATTA, 'Punjabi Refugees and the Urban Development of Greater Delhi', in R.E. FRYKENBERG op.cit., pp. 442-462

40 Since 1975 there have been repeated expressions of concern about the future as well as the past of Shâhjahânâbâd, about rehabilitation of those living in very congested mahallahs, as well as conservation of its cityscape. At least four workshops were held to discuss these issues between 1980 and 1988, the first of which was sponsored by the School of Planning, Delhi, and the Max-Mueller-Bhavan. This was later published at Shâhjahânâbâd - Improvement of Living Conditions in Traditional Housing Areas ed. by B. GHOSH (Delhi 1980)

through lanes and streets which are the veins on the leaf of the original city, but the leaf has become dessicated the green of the gardens, the sparkling water in the canal, the slow pace of the horse-drawn vehicles, the low decibel-level, the graciousness which grows in an old-established community, these are lost. The Taj Mahal still has much of the beauty that Emperor Shâhjahân gave it, but his city has died many deaths. Nâdir Shâh stripped it of its wealth, the British distorted its form with partition much of its cosmopolitan culture disappeared, but the city is alive, and has a more human face than many of the stately and spacious streets of New Delhi.

Islamic Institutions and Infrastructure in Shâhjahânâbâd

Jamal Malik (Heidelberg/Bonn)

Preliminary remarks

The map under consideration here bears inscriptions in Urdu, a language of the Indian subcontinent. Even though the anonymous cartographer mastered the writing of the Urdu script, there are a number of orthographic mistakes which have hindered the unequivocal identification of quite a number of streets, buildings and names. The errors committed seem to be typical for a person of native Hindi tongue, especially when it comes to proper names. The absence of many important names and especially of the Persian grammatical classifications may indicate that the map was drawn up in considerable haste. One may thus suppose that the cartographer was of Hindi background and that he served in the colonial administration.[1] Moreover, the indications made by him foster the overall impression that Shâhjahânâbâd was an almost exclusively Muslim city; centres of Hindu and popular religious discourse like dharamsalas or temples and especially shrines can hardly be located. This would indicate that the cartographer might have been very particular about Muslim institutions for a variety of—political—reasons. This assumption may, however, be undermined by the facts that even non-Muslims affiliated to the court at that time were usually conversant with Persian and that several important Muslim buildings are not given on the map.[2]

---

1 Some errors may already be pointed out here, the cartographer's orthography followed by the correct one in parantheses: Kûchah Kâbil 'Attâr (Kûchah-e Qâbil 'Aṭṭâr), Khasânah (Khazânah), Suman Burj (Muthamman Burj), Nâz (Naẓârat), Ḥawelî Nâz (Ḥawelî Naẓârat), Ahâtah Kâgadhî (Aḥâṭah Kâghadhî), Kûchah 'Akil Khân (Kûchah-e 'Aqil Khân), Ḥawelî 'Aẓim Khân (Ḥawelî A'ẓim Khân), Kûchah Kahan Chand (Kûchah-e Khân Chand), Ummîd al-Dîn ('Ummîd al-Dîn), Kamar al-Dîn Khân (Qamar al-Dîn Khân), Kûchah Ruh Ullâh Khân (Kûchah-e Rûḥ Ullâh Khân), etc., as well as the words ḥawz (ḥawḍ) and hâtâ (ḥâṭah). The Persian particle *-e*, which signifies the possessive case, is generally missing. The cartographer preferred, however, the Arabic-Persian orthography of *la'l* to the Hindi version *lâl*. There is no denying the fact that while in identifying, reading and transcribing the difficult entries on the map, several mistakes and errors may have been made. Suggestions for corrections and improvements will therefore be appreciated and are most welcome.

2 Some of these missing institutions have been marked in this contribution by reference to "(near nn)", indicating their approximate locations. Especially the proper names of smaller mosques have been identified by contrasting the map of around 1850 with the one of 1879 as elaborated

The paper in hand deals with the topography of Shâhjahânâbâd as it appears in the map of around 1850 and with reference to the religious—Islamic—institutions. Starting out with a first general glance at the spatial order, focus is then laid on the different quarters, the *maḥallah*s, which often bear the suffix *-wârâ*, *wârah* or *-wâlâ*. This basic pattern is then examined in detail with reference to Islamic institutions and their specific social environment. The significance and functions of religious scholars and sufis is discussed in the context of urban quarters, and their respective spheres of influence are identified. The tradition of the mystic Naqshbandi order and its ramifications will serve as an example.

Spatial order

Shâhjahânâbâd was a manifestation of the imperial rule of the Mughal emperor Shâhjahân (died 1666), successor of Jahângîr (died 1627), and predecessor of the last great Mughal, Aurangzeb (died 1707). By designing this town according to the pattern revealed on this map, the builders of Shâhjahânâbâd created the architectonic expression of what has often been called the "patrimonial system" in its climax. According to tradition, the emperor soon invested the Islamic gentry (*shurafâ*) with land and tax exemption (*madad-e ma`âsh*). The *shurafâ* themselves originated mostly from the *qaṣbah*s, garrison posts and administrative settlements in which Islamic scholars also met their clients and where an integrative or even syncretist culture prevailed—usually established around a tomb and a *waqf*, pl. *awqâf* (property which cannot be transferred and which is therefore inalienable; to a certain degree also *res extra commercium*), which still has important spatial-, socio-economic and political functions (Ehlers 1991, pp. 100-103; Malik 1990). The *qaṣbah*s scattered in the *hinterland* not only provided the imperial city with necessary manpower and financial resources but were also recipients of imperial benefits, and some of them developed into centres well known for their particular functions—trade, artisanry, or scholarship (Bayly; Alam). Thus, the local potentates in the *qaṣbah*s were able to gain a certain degree of independence or even autonomy which led, among other factors, to the "crisis" of the Mughal Empire (Alam). In this context, the term rent-capitalism—its parasitical and exploitive character with regard to the urban culture's rural *hinterlands* (Ehlers 1978)—and the notion of "central place system of dominance" (Ehlers 1991, pp. 96f) may probably be re-thought for the region of South Asia. It is true that religion has "greatly contributed to the social, and economic supremacy of the cities over their rural hinterlands and to the overall positive image of urban life in comparison to non-urban life styles" (Ehlers 1991, p. 103); however, this seems to be a rather recent phenomenon since scholars of the 16th and 17th centuries turned to *qaṣbahs* which

...
by Aḥmad. In addition, religious—Islamic—institutions may be found on the map "Islamic Infrastructure" which is included here.

again regained their connotation as proud *waṭan* (lit. homeland) in the wake of colonial penetration. In fact, the Islamic scholarship which flourished in the *qaṣbahs*, reflected a high degree of normative independence from the cultural and imperial metropolis (Malik 1992).

In Shâhjahânâbâd the feudal tenures of the *shurafâ* usually were situated to the west of the palace, along one of the two boulevards—at the Chândnî Chawk—, and, originating from the emperor's palace, thus furnishing the city with an unequivocal structure. Other social groups hierarchically settled around the palace according to their respective social status. Important markets (*bâzâr*s) soon developed along the connecting lines of important institutions like the Fataḥpûrî Masjid, the Qâḍî kâ Ḥawḍ (cistern of the qazi), the Jâmi' Masjid and the Kalân Masjid. Smaller markets were able to establish themselves in the areas to the south and north of Chândnî Chawk as they were related to the groups settled in this area. Some space close to the city walls remained open as working place for the lower strata of society (e.g., sweepers, potters and leather workers). In the 18th century, one could subdivide Shâhjahânâbâd into three rough categories:

1. North of Chândnî Chawk the area remained in possession of the gentry with its mansions, gardens and palaces.
2. Christian missionaries and European traders settled in Daryâganj (in the southeast). After the topographical changes in the wake of annexation of Shâjahânâbâd in 1803, the Europeans also managed to gain a foothold in other areas, e.g., in the region surrounding St. James' Church and near the Kashmiri Gate (see Gupta 1981, p. 16).
3. The majority of the population lived and worked south of Chândnî Chawk (see Fonseca, pp. 72ff), e.g., in Galî Rodgarân (gut-workers) (west of Qâḍî kâ Ḥawḍ), while the poorer strata, such as the kumhâr (potter), qasâ'î (butcher), dhobî (washer), chamâr (tanner) and telî (oil-extractor), etc., predominantly lived close to the city gates or even outside the city wall, with the exception, however, of the Lahori Gate, the Kabuli Gate and the Kashmiri Gate as well as the eastern entrances to the city (see Gupta 1981, pp. 53-55; Gupta 1986, p. 255).

Also, the city can roughly be subdivided according to the representatives of the two dominant religions, the Hindus and the Muslims: While Hindus predominantly lived in Chhîpîwâṛâ (cloth-printers) (west of Jâmi' Masjid) and in North-Billîmarân (southeast of Fataḥpûrî Masjid), the majority of the Muslims were settled in South-Billîmarân, Lâl Kû'ân (red well), Ḥawelî Ḥaider Qûlî Khân, and close to the large mosques.

The Maḥallah

A more precise spatial subdivision can be envisaged when referring to the particular *maḥallah*s, which form a pattern of differenciated quarters. The quarters are embedded in a complex texture with their norms relating not only to economic necessities but also to manifold social interweaving. They are socially cohesive (Chandhoke, p. 14; Fonseca, p. 75; Gupta 1981, p. 2), since prior to the structural upheaval under colonial rule there was no separation of the spheres of production and reproduction. Consequently, the quarters mostly bear the denomination of the dominant service sector settled there, i.e. artisans, traders, ethnic groups or other representatives of economic or social life, e.g., the Maḥallah-e Dhobiyân (washermen's quarter), Maḥallah-e Sawdâgar (traders' quarter) (near Jhajjar wâle Nawwâb), Maḥallah-e Muftiyân (quarter of religious scholars) (near Muftî Ikrâm ad-Dîn), Maḥallah-e Teliyân (oil-extractors' quarter) (near Shîsh Mahall), Maḥallah-e Rikkâb (quarter of either stirrup-holders or cupbearers), Maḥallah-e Sû'î wâlân (quarter of those producing needles), Maḥallah-e Punjâbî (Punjabis' quarter), Maḥallah-e Ḍadariyân (shepherds' quarter) (near Kûchah-e Shâh Turkmân), Katrah-e Mârwâṛî (quarter of those belonging to Marwar), Jatwâṛâ (quarter of the Jats).

The local representatives of the different social and ethnic groups aligned their buildings and the adjoining streets in a functional manner. Inside a *maḥallah*, often comprising a *katrah*—an emporium also offering lodging—as a centre, small alleys (*galî* or *khûchah*) can be found; they also usually bear the names of the corresponding professions or ethnic groups, e.g., Imâm kî Galî (preachers' lane), Galî Gur wâlon kî (lane of those producing or selling raw sugar, jaggery), Gandhî Galî (lane of flower vendors), Râmpûriyon kî Galî (lane of people belonging to Rampur). The greater the distance to the core of the *maḥallah*, the broader and the more representative—but also increasingly socially anonymous—are the lines of communication, which may be categorized as "primary, secondary and tertiary streets" (Fonseca, p. 74). Only in exceptional cases was a *maḥallah* the exclusive homestead of just one group or profession: the Mahallah-e Rodgarân (quarter of the gut-workers) in the west, e.g., or the Mâlîwâŕâ (gardeners' quarter) in an easterly direction gave shelter to different groups of artisans in just one quarter. Therefore, an unequivocal social and professional classification is not possible through reference to the *maḥallah*s alone. Besides the economic institutions mentioned here, the important and identity-giving religious institutions may serve as a useful means for locating specific professions since these institutions determine the life inside as well as outside the *maḥallah*.

Religious institutions as signposts of the social fabric

We may gain detailed insight into the structure of the city by locating the places of religious discourse—mosques, shrines, tombs and temples—and by observing their

Islamic Institutions in Shāhjahānābād around 1850

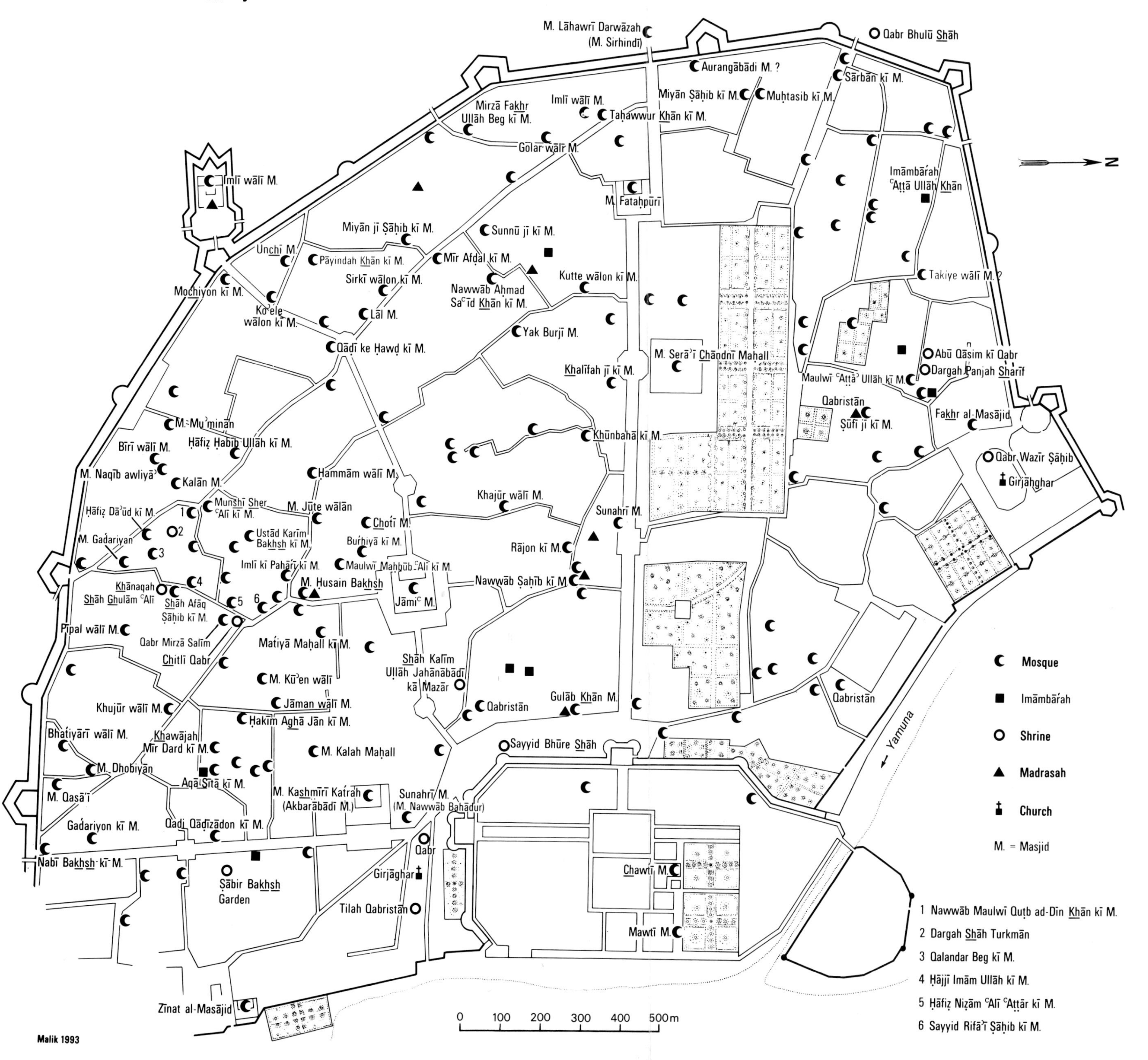

relations to the alleys and the main lines of communication within the city. Mosques in Shâhjahânâbâd certainly had a pivotal role: According to a decree of Shâhjahân, "in every lane, bazar, square, and street" a mosque was to be founded (Blake, p. 181), most of them being religious endowments or *waqf* (Sanderson).[3] Mosques are the material expression of an exclusive Islamic culture—the incontestable symbol for the teachings of Muhammad—dominating the local cultures. According to Blake, p. 179, in Shâhjahânâbâd they were "more numerous and significant than *dargahs*, *khanqahs*, *imambarahs*, or *idgahs*", the centres of popular religion which were mostly *awqâf* as well. However, especially in the Indian environment, the message of Islam could only develop through Sufism and its institutional manifestations, the holy man and his *dargâh* and *khânaqah*: "Indian Islam seems to have been essentially a holy-man Islam" (Trimingham, p. 22; see also the contributions in Troll 1989), and the *dargâh*s were most important even for the urban Muslims who had by and large come from the *qaṣbah*s, where the cult around holy men and tombs usually determined the identity of these towns, in an environment which was dominated by Hinduism. Later, it was the *khânaqah*s and *dargâh*s of the Naqshbandiya-Mujaddidiya order or the *imâmbârah*s of Shiites which became main centres of political and religious discourse.

These numerous institutions played a central role for the integration of representatives of different social and regional origins as well as political, economic and religious affiliations. In contrast to the exclusive mosques, they are the material expression of a "melting pot" kind of *weltanschauung*; they are microcosms of an indigenous Islamic culture with regular celebrations and rites. In Delhi a discernible shrine and popular culture was established rather soon (its inventory up to the mid-18th century is given, e.g., in Ḥabîb Ullâh). The most notorious shrines, to be sure, are located outside the city, but even inside of Shâhjahânâbâd the number of small *dargâh*s numbers several hundred. The cartographer, however, has hardly indicated any of these widespread centres of popular culture. Only the *dargâh*s—all *waqf*—of Sayyid Bhure Shâh, of the Chishti Sufi Sayyid Ṣâbir `Alî alias Ṣâbir Bakhsh (died 1821), and of Shâh Turkmân (died 1249) are listed in the map. The latter was a well-known Sufi affiliated with the Suhrawardi mystic order. The quarter here is called Maḥallah-e Qabristân due to its numerous mosques and tombs (see Khân, III, pp. 333f; Aḥmad, pp. 156ff, 309; Sanderson, pp. 55ff, nos. 99ff). Likewise, the important Shiite Dargâh Panjah Sharîf near Gandah Nâlah (near Biṛ wâlî Masjid) is not indicated by the cartographer. Therefore, the map does not enable us to reconstruct the location of mystics, who also represent certain professional groups.[4] Much in contrast to this lamentable omission,

3 Sanderson has, among other things, given a descriptive list of mosques, indicating their status mostly as *waqf*. His elaborations seem to have been based on the more detailed information provided by Aḥmad and the map of 1879.

4 A connection between mystics and handicrafts, that is mystic orders and guilds, can be recorded in the Islamic Middle East. A study which would elaborate on this correlation for the region of South Asia would be most welcome.

we find a host of mosques on the map, permitting us to rely on this method. For example, those professional groups delivering fresh agrarian products to the city (i.e., shepherds and butchers) who were in need of a corresponding quick access to these goods obviously must have settled along the southern and south-south-western rim of the city walls (Dehli[5] Gate and Turkman Gate): this is where their ritual institutions, such as, e.g., Masjid Gaḍariyon (shepherds' mosque), Masjid Qasâ'î or Masjid Qaṣṣâb (butchers' mosque) were located. The washermen also used to live in this area, as indicated by the Masjid Dhobiyân (near La'l Miyân kâ Chhattâ). In the south-western part of the city, the narrow lanes and mosques of the rough potters, metal workers, masons, the coalers, etc. can be found. They all represent "low ranking" trades.

The closer to the core of the city the more socially recognized are the professions settled there: weavers, producers of wool, traders of saddle-horses, oil-extractors and manufacturers of straw goods, each of them represented by their respective mosques. Along the "primary streets" we even find the caretakers of central institutions, such as, e.g., the collector of taxes—the Karoṛî (for his role see Richard, pp. 39f). The Muḥtasib (near Rang Maḥall), i.e., the censor of public morals, esp. in the market area, is also represented by a mosque (for the role of the institution of the Muḥtasib, which was reintroduced under Aurangzeb in 1659, see Siddiqi).

Those professional groups further processing the different goods as well as those of the public service sector are situated north of the axis Sîtârâm Bâzâr or Kûchah-e Sayyid Qâsim Khân. Here we encounter carpenters, dyers, gardeners, shoemakers, potters, money-changers, manufacturers of bangles, spinners, basket-makers, even hair-dressers, etc. Each of these groups had its own little mosque, in which it practised its respective rituals and where the local religious scholar or in many cases the local hereditary holy man practised the *rites de passage*.

Further, in the direction of Chândnî Chawk, mostly representatives of the trading professions, e.g., traders of fabrics, fish, meats, luxury goods, but also some of the professional groups processing goods, e.g., producers of waterpipes (*ḥuqqah*) can be found. All of them are characterized by the spatial proximity to the imperial house. They were therefore represented by the respective religious dignitaries legitimizing imperial rule.

Along the street at Chândnî Chawk, the luxury shops are located where one could purchase the best of the ready-made goods. The representatives of these professions were symbolically represented by the large and magnificent mosques, e.g., the Sunharî Masjid. At the end of Chândnî Chawk, we find the Fataḥpûrî Masjid. In a western-north-western direction from this point there are some *maḥallah*s around Khâṛî Bâ'olî; they have specialized in trade and refining relatively high quality goods such as, e.g., tobacco, flowers, perfumes, butter-oil, pomegranates. It is in this economically well-to-do region where the mosque of the Muḥtasib is located. To the north of this

5 For Delhi, the cartographer has used the version as prevalent among Muslims, i.e. Dehli.

important area of access the well-known Punjâbî Katrah is located, populated by the ambitious and influential traders and workers who specialized in many different professions (Ṣâbrî, p. 50) and who engaged in building the respective mosques (similar activities can be noted for the Jain community with regard to Dharamsalas und Pathsalas; see Gupta 1981, p. 54). Dancing girls as well lived in this area (Galî Kânchne kî).

At the periphery of *maḥallah*s, usually marketplaces and the working spaces (*katrah*) of the Dhobîs are to be found. This widespread professional group also has a *maḥallah* to itself and a ritual centre. Professional groups on the lower ends of the social ladder, however, seemingly did not have any proper mosque, even though they were spread all over the city; this is the case, e.g., with the Muslim tanners.[6] In the same way it is not possible to locate the widespread barbers on the map. The map by Aḥmad dated 1879, however, points to a *maḥallah* called Nâ'îwâṛâ (near Shâh Bolâ kâ Baṛ).

Prior to the introduction of European ready-made goods, the different regions of the city had several local indigenous manufactories, *kârkhânah*s, established under the patronage of the *shurafâ*, connected with religious institutions (see Gupta 1981, p. 52),[7] and located about town according to the type of product they produced. From the outskirts of the city towards the centre, a specialization pointing to the hierarchical character is discernible: the manufactories situated northeast of Sîtârâm, between Turkman Gate and Lahori Gate, delivered goods to the workshops further northeast. The *kârkhânah*s at the outer rim received the processed raw materials from the Maḥallah-e Rodgarân, Maḥallah-e Namde wâlâ (people who produce and sell woollen cloth being made without weaving) (near Maḥallah Kaṭne kâ), etc. and refined them. In the direction of downtown, there were more manufactories, characterized by their ever increasing quality of goods (Ṣâbrî, pp. 35-38).[8] No data is available yet on the

6 The question of why tanners were spread all over the city remains unanswered. Perhaps it was because of the high demand of the army.

7 The connection of religious institutions and artisanry and handicraft supports the thesis that even in Shâhjahânâbâd, similar transactions and correlations existed as were prevalent in the *qaṣbah*s.

8 South, next to Turkman Gate, there used to exist the *kârkhânah* called Urdû Mal Paimak wâlâ, which produced gold or silver brocade, most probably for the military (urdû). Next to this institution, in Maḥallah-e Sû'î wâlâ near the Ḥawḍ Mîr Khân, metals were processed in the manufactory of a certain Khalîfa Zafar. To the west, in Dargâh Shâh Turkmân, silk was refined at Mirzâ Walî Beg Baṭṭiyâ. To the north, in the direction of Jâmi' Masjid, at the Imlî kî Pahâṛî, there was another manufactory specialized in refining precious metals called Amîr kâ kârkhânah. Here one would also find Badr ad-Dîn Khân, who was famous for edging. In Maḥallah-e Garhiyâ—west of the metal and lace processors—was the *kârkhânah* of 'Abd al-Karîm Zardawz; when the number of employees increased here, this *kârkhânah* was shifted to Chândnî Chawk. Near the southern entrance of Jâmi' Masjid, in Kûchah-e Matiyâ Mahall—which seems to have served as a temporary residence for Shâhjahân while the fortress was being built and which later was given to a harem lady called Nawwâb 'Azîzâbâdî Begum—was the brocade manufactory of Mîr Ṣafdar 'Alî Goṭe wâle, who formerly lived in Matiyâ Mahall. Next to this, at the Bâzâr Chitlî

religious institutions frequented by the laborers engaged in the *kârkhânahs*; it may be presumed, however, that all these manufactories reflected specific common ritual communities and thus had their own particular mosques/shrines.

The spatial distribution of mosques marked out here supports the rough spatial order of the city noted above. It may be witnessed from south to north as well as from west to east: The poor strata usually lived close to the city walls (esp. at the Dehli Gate, the Turkman Gate and at the Ajmeri Gate in the south, Farâshkhânah in the west and Mori Gate in the northwest) (see also Gupta 1981, pp. 54ff).[9] Their neighbors were the processing professions and tradesmen, and finally came the rich high-ranking strata at Chândnî Chawk and at its *bâzârs*.

Hierarchy of mosques

This internal hierarchy was part of the concept of the city, a fact which becomes evident from the emperor's allocation of land to the *shurafâ*. It was repeated in the decree of Shâhjahân concerning the construction of mosques primarily from the east to the west, or north to south, following the imperial perspective. This official city planning necessarily had to manifest itself in a hierarchy of mosques itself: The royal (*pâdshâhî*) Jâmi' Masjid was succeeded by eight elite (*begumî—amîrî*) mosques constructed by notables. These were the Masjid Fataḥpûrî, Masjid Akbarâbâdî, Masjid Sirhindî, Masjid Aurangâbâdî, Zînat al-Masâjid, Sunharî Masjid, Masjid Sharîf ad-Dawlah and Fakhr al-Masjid. They were all built between 1650 and 1728 (see Blake, pp. 182f). They were located next to the two main lines of communication which subdivided the city. The elite mosques seem to have "attracted a cross-section of persons from all over the city. As with the Jami Masjid, worshippers came from no special group or class." (Blake, p. 183).[10] At the other end of this hierarchy stood the

...

Qabr—the name Chitlî Qabr is based on the coloured tomb of a holy man who lived here at the end of the 14th century; in this *maḥallah* a number of tombs are situated—in the Mîr Ḥâshim ke Phâṫak another brocade manufactory flourished. It was known as Nûr Muḥammad Tâsh wâle. Finally, at the Kûchah-e Khân Chand next to Chândnî Chawk, the *kârkhânah* of Ustâd Amîr was established; it specialized in trimming lace of gold or silver. In the northern part of the city, to the east of Punjâbî Katrah, at the Maḥallah-e Gandâ Nâlah near Kashmiri Gate, the factory of Muḥammad Jân Les wâle was situated and he was known as the only producer of adhevise plaster in town. Westwards of the Fataḥpûrî Masjid at Phâṫak Ḥabash Khân one could find the manufactory of Amîr Bakhsh Zardawz. In accordance with local demand, this institution specialized in the outfitting of cavalrymen, and courtly dress. Some way to the southeast, at the Farâshkhânah, the large enterprise of Muḥammad Raḍâ Goṫe wâle was located; it again produced brocade. These manufactories could, according to their size, employ 20 to 500 workers, in some cases exclusively women.

9 Also in the northwest probably the Pâgalkhânah, i.e. the lunatic asylum, was located.

10 As will be shown later, this statement may be modified, for later some elite mosques, such as the Akbarâbâdî Masjid and Fataḥpûrî Masjid, became centres of particular schools of thought and sought their substratum among particular social and professional groups.

so called *maḥallah*-mosques, numbering about 200 in Shâhjahânâbâd. A large part of them was also constructed by high-ranking Mughal officers, influential traders and religious scholars and sometimes located within the *ḥawelî* (mansion) of a member of the *shurafâ* (see Blake, pp. 183f). One may assume that the founders of these mosques were associated with one or another professional group which worked and lived there. These mosques often bear the name of their founders and they are mostly located on "secondary streets".[11]

Some of the mosques of higher social ranking are named after the professions active in these localities, e.g., Qâḍîzâdon kî Masjid (mosque of the qazis) or Muḥtasib kî Masjid (near Rang Maḥall), and they are mostly located at the important —"secondary"—lines of communication. Besides having representational purposes, they certainly have the function of addressing larger masses, in keeping with the statutes of Islamic Law (see Blake, p. 84). This perception was of special significance for the mosques around the Bâzâr Chitlî Qabr and those close to the Akbarâbâdî Masjid. The Naqshbandi Sufis who had settled there were able to develop these institutions into centres of their propaganda and mission due to several factors, their relatively favorable locations not the least among them (see below).

Another part of the so called *maḥallah*-mosques were those built and financed by immigrant groups, e.g., in the case of the Punjabis who had mainly Hanafite affiliations (Masjid Punjâbiyân) and used to live near the Kabuli Gate and within the Punjâbî Katrah. In accordance with this, the elite mosque Akbarâbâdî Masjid became known as the Masjid Kashmîrî Katrah, as Kashmiri traders had settled near it.

Artisans, laborers and vendors living along the city walls in the south and in the west agglomerated in the so called "tertiary streets" with their respective mosques.[12] In rare cases, the *maḥallah*-mosques were named according to historic events, e.g., the Shiite Khûnbahâ kî Masjid.

---

11 E.g., Masjid Raushan ad-Dawlah, Nawwâb Sâhib kî Masjid, Mîr Dard kî Masjid (near Kûchah Chelon kâ), Sayyid Rifâ'î Ṣâḥib kî Masjid, Maulwî 'Aṭṭâ' Ullâh kî Masjid.

12 E.g., Kahâr wâlî Masjid (water-bearers' mosque) (near Kûchah Chelon kâ), Dâ'î wâlî Masjid (nurses' mosque) (near Tirâhah Bairâm Khân), Gawndnî wâlî Masjid (gum-fruit tree's mosque) (near Ḥawelî Khân Dorân Khân), Bhaṫiyârî wâlî Masjid (innkeepers' mosque) (near Masjid Qasâ'î), Pulâ'o wâlî Masjid (mosque of the rice-dish) (near Dargâh Shâh Turkmân), Masjid Jûte wâlân (shoemakers' mosque) (near Hawelî Sîtalwâṙah), Sirkî wâlon kî Masjid (reed-grass workers' mosque), Anâr wâlî Masjid (mosque of the pomegranate) (near Galî Hingâ Beg Khân), Masjid Ko'ele wâlon kî (coalers' mosque), Râjon kî Masjid (masons' mosque) (near Galî Pîpal wâlî), Masjid Gadariyon (sherperds' mosque) (near Jaṫwârâ), Masjid Qasa'î (butchers' mosque), Masjid Dhobiyan (washermen's mosque) (near L'al Miyân kâ Chhattâ), Khajûr wâlî Masjid (mosque of the date tree) (near Kûchah-e Rûḥ Ullâh Khân), Golar wâlî Masjid (mosque of the wild fig tree) (near Samosiyon kî Galî), Baṙ wâlî Masjid (mosque of the banyan tree), Pîpal wâlî Masjid (mosque of the fig). The location in the centre of the city—vital to the fish vendors—enabled them to build a mosque in this place, the Machlî wâlon kî Masjid within the Chitlî Qabr Bâzâr.

The map further shows that one *maḥallah* may frequently shelter several professions, which then constitute a respective sub-*maḥallah* with its formal manifestation and delimitation marked by respective alleys and by its symbol, a mosque. Thus, the assignment of a mosque to a specific *maḥallah* is quite problematic. In this context it therefore seems more advisable to use the term "profession-mosque".

The localization of mosques reveals a vivid picture of religious institutions, density increasing with the growing number of the groups of traders, vendors, artisans, and workers. The majority of the "profession-mosques" is therefore located south of Jâmi` Masjid, an area occupied by those professions supplying the city with fresh natural products and with consumer commodities. A similar concentration of mosques—though less dense—can be located along the western and the northern city walls.

The culture of religious scholars

The question of how the upholders of culture, the religious scholars, were integrated into the complex social fabric, is crucial for understanding the interaction of different groups in the city. What was their role in the social dynamics, in spite of—or perhaps even because of—the hierarchical structure of the society, reflected in the city's topography? For many centuries, Delhi has been a stronghold of Islamic culture and sciences, a fact sufficiently known. This is not the proper occasion to deal with the origins of Islamic scholarship in India. Suffice it to say that even prior to the rise to power of the Mughal emperors (i.e. from 1526), the then emperors employed religious scholars and mystics from western adjoining regions—esp. from Persia—at their courts, thus supporting a Persian-Shiite scholarship. This tradition was continued by the Mughals, who generously invested persons with land. Simultaneously, the annual pilgrimages to the Hijaz supported communication among religious scholars. As a result, Delhi soon became renowned for its studies in Hadith and in mystic theosophy.

One well known mystic tradition in the region of Northern India was that of the Chishti order, which spread rapidly due to its integrative and liberal teachings, thus sparing the inhabitants of the city from sectarianism until the first quarter of the 18th century (see Chandra, pp. 210ff). The Chishti tradition stood in some opposition to the Naqshbandiya order, which gained in strength from the 17th century onward. This order soon demanded a reform of sufism and consequently sometimes isolated itself from the popular religious environment.

The religious scholars and sufis unfolding their activities here have not only left behind numerous sacred institutions, they must also have represented different social groups in several ways. Some of them were active in the service of the emperor, thereby mediating between the court and the social base. Others rejected any cooperation with the ruling strata, isolating themselves from secular power and thereby appealing to marginalized social groups. Between these two poles there existed a multitude of other Islamic cultures, their different world views often overlapping one another. In the following, we will deal with these supporters of Islamic culture and will try to demonstrate how the city is subdivided into different Islamic groupings, which occupied certain areas due to their specific social background. In this way the dominant expressions of Islam in South Asia can be identified on the micro-level. Focus will be laid on the tradition which had the strongest influence on Delhi in the 18th und 19th centuries: the order of the Naqshbandiya and its branches.

The Naqshbandiya

One outstanding personality of the Naqshbandi tradition was Aḥmad Sirhindî (1563-1624), disciple of Khwâjah Bâqî Bî'llâh (1563-1603) (see Algar, pp. 19ff; Rizvi 1990, p. 157 et passim).[13] Among other issues, Sirhindî propagated the purification of Islam from Hindu elements as well as from those elements typical of popular belief. His ideas thus seem to have been not only in contrast to the indigenous integrative, sometimes even syncretist culture; they were also in contrast to the ideas of the Shia.[14] His doctrines were supported in Delhi by three mystics in particular: Shâh Walî Ullâh (died 1763), Mîr Dard (died 1785) and Mirzâ Maẓhar Jân-e Jahân (died 1781). While some of the traditionalists tried to unite the Muslims and to integrate them for common political action (this was most important due to the invasion of the Marathas), others rigorously tried to enforce Sirhindî's purist approach. This may have led to an increase of internal Islamic disputes, the first climax of which can be witnessed in the confrontations between Sunnis and Shiites in the 18th century (see f.e. Qâdirî).

The reformer Shâh Walî Ullâh traced his spiritual descent back to Aḥmad Sirhindî, as did most of the Naqshbandis. He had studied with, among others, his father, Shâh 'Abd ar-Raḥîm (died 1718), who had established the Madrasah Raḥîmiyah in Mahndiyân outside the city walls. When Shâh Walî Ullâh returned from the Hijaz in 1732, he integrated the books on Hadith into the secular-rationalist (*ma'qûlât*) *dars-e niẓâmî*, which had been developed by religious scholars in Lucknow, and he re-established the traditional sciences (*manqûlât*). Specializing in *manqûlât*, the Madrasah Raḥîmiyah now attracted many students, so that the Mughal emperor Muḥammad Shâh invested the founder of this school with the Ḥawelî Kalân Maḥall in Daryâganj (near Faiḍ Bâzâr) in the form of a *waqf*. In recognition of his outstanding scholarship, Shâh Walî Ullâh was additionally awarded 51 bigha[15] of land near Delhi in 1754 by 'Alamgîr II (reigned 1754-1759) for the construction of his own *madrasah* (see Qâsimî, pp. 42-48; also Rizvi 1982, pp. 84f).

13 Why Bâqî Bî'llâh succeeded in asserting his ideas during his short four-year stay in Delhi is still as unanswered as the pertinent question about the social base of the order, not only in South Asia.

14 Aḥmad Sirhindî's aversion to the Shia may have been based in the fact that the Naqshbandiya was persecuted in the Shiite Safavid empire.

15 A *bigha* was the common land measure in Mughal times. It was, however, subject to regional variation. In Northern India one *bigha* measured approximately 3000 square yards; see Abu al-Fazl: *A'in-e Akbari*, Vol. II, pp. 62f.

Shâh 'Abd al-'Azîz and his mission

Due to various circumstances, Shâh 'Abd al-'Azîz (1746-1824) and his brothers[16] considered themselves called upon to take up the succession of their father, Shâh Walî Ullâh: Three important representatives of local religious culture had died towards the end of 1785: Maẓhar Jân-e Jahân, Mîr Dard and the Chishti mystic Fakhr ad-Dîn (died 1785). Furthermore, the Sikhs had attacked Delhi frequently and, finally, the Shiite Mirzâ Najaf Khân (1772-1782), who had come from Kabul to the capital at the time when Shâh 'Alam II (reigned 1759-1806) returned to Delhi, was making life difficult for the Sunnis (see Rizvi 1982, p. 22-29 et passim).

As religious reformers the brothers tried to put an end to the "un-Islamic practices". They held the Shia responsible for political and moral decay, esp. referring to the Shiite practice of *taqîya*, which allowed any Shia to deny his religious identity in certain cases (Rizvi 1982, passim). Shâh `Abd al-`Azîz's postulates of reform appealed rather to the outer, politically dominant public and were focused on the structures which were gradually undergoing change in the course of colonial penetration. His intentions were reenforced by the British attitudes about becoming involved in local cultural affairs after 1803: this attitude had disqualified local rites as an expression of a decaying culture as well (see Fusfeld, pp. 40f).[17] In order to counter this colonial critique, Shâh 'Abd al-'Azîz composed a large number of formal legal opinions (*fatâwâ*); they were, however, limited in character as integrationist attempts that aimed at providing the Muslim community with a legal basis for action in an Islamic Empire which seemed to have been lost to the British once and for all.[18] He nevertheless cultivated his relationship with the colonial public and with other social groups, or "intellectual circles of Delhi" (Damrel, p. 276)[19] which had access to the British administration. Thanks to his good relations with the colonial masters, administrative authorities awarded to him his father's landed property shortly after the annexation of

16 Other sons of Shâh Walî Ullâh were Shâh Rafî' ad-Dîn (1749-1817), Shâh 'Abd al-Qâdir (1753-1827) and Shâh 'Abd al-Ghanî (1758-1789).

17 Before and especially after the annexation, polemic treatises by and disputes between Christian missionaries and Islamic scholars were *en vogue*. Before 1857 most of them were carried out in peaceful manner, sometimes in mosques.

18 These *fatâwâ*—which before the colonial critique consisted of questions and answers only without reference to sources—could now stand up to the Western "academic" and "scientific" demands.

19 Shâh 'Abd al-'Azîz's strong affiliations with the imperial sector may also have been one reason for his limited interest in mysticism. However, his brothers and followers later tried to address the masses, who usually were inclined towards mystic tradition, thus making accessible the reformist programmatics through Quranic translations and the new media. The application of colonial media culture by Islamic scholars presupposed, of course, that those in professions highly recognized before the entry of the colonial masters, like paper producers, writers and printers, represented a part of the social base of the movement and through the Naqshbandi propaganda saw a way to preserve their status.

Delhi in 1806; his father had been invested with this land by 'Alamgîr II and it had been withdrawn from him in 1774 under the influence of the Shiites (see Fusfeld, pp. 27f, Rizvi 1982, pp. 84-91; the detailed correspondence between Shâh Abd al-'Aziz and the British administration concerning this landed property is documented in Rizvi 1983). Shâh 'Abd al-'Azîz then established the Madrasah Shâh 'Abd al-'Azîz, on a site located between Chitlî Qabr and Tirâhâh Bairam Khân.[20] Until his death, he held sermons every Tuesday and Saturday with an audiance of several thousand followers (Rizvi 1982, p. 93)—most probably in connection with local markets. The compound, a *ḥawelî*, offered sufficient space for the accomodation of teachers and students. After 1857, however, this school was destroyed by the British, just like the Akbarâbâdî Masjid. His brother, Rafî' ad-Dîn, teacher in his mosque in Chitlî Qabr, supported him in his Madrasah. His other brother, 'Abd al-Qâdir, was also considered one of the recognized religious scholars of Dehli and was engaged in teaching in the large Akbarâbâdî Masjid (see Qâsimî, passim; Ahmad, pp. 173f). It was here that leading activists such as Shâh Ismâ'îl (1793-1823) and Shâh Aḥmad Shahîd gave their lessons.

Shâh Aḥmad Shahîd (1786-1831), a soldier by profession and a mystic, originated from a *qaṣbah* near Lucknow, later studied under the sons of Shâh Walî Ullâh and soon became a leading personality. The abolition of the cult of saints was also the focus of his endeavours, which, after a phase of dispute, finally became a militant movement of the *Mujâhidîn*. Delhi became its center of recruitment and its organisational area, with the political and theological centre being the Akbarâbâdî Masjid, which maintained an animated exchange with the Madrasah Shâh 'Abd al-'Azîz (Rizvi 1982, passim; Qâsimî, pp. 56f).[21] The bustling trade at this mosque, its relation to the Kashmiris (thus the name Masjid Kashmîrî Katrah), the good conditions of communication and the proximity to different markets bordering the Jâmi' Masjid as well as to the manorial area were important features for the reformist movement in fostering its influence among specific social groups represented in these rather well-to-do areas.

Naqshbandiya-Mujaddidiya

A different tradition of the Naqshbandiya evolved with the person of Mirzâ Mazhar Jân-e Jahân, who established a split-off of the order, the Naqshbandiya-Mujaddidiya. His considerations for reform had also been formed by the decay of the Mughal empire, the intrigues at the court and the violant clashes which he always pointed out to his group (*ḥalqah*) in the Bâzâr Chitlî Qabr. He thus considered himself as well as his doctrines to be an alternative to the contemporary collapse and soon put aside his family contacts with the court and increasingly withdrew from worldly affairs,

---

20 On the present map, this important *madrasah* is, however, not indicated.

21 Of course there was also strong opposition against this movement in Delhi. Shâh Aḥmad Shahîd and his followers soon migrated to the northern part of India, to the area of tribal society, but they were not able to establish a durable improvement for the Muslim community.

adopting an attitude which is commonly labled as quietistic (see Fusfeld, pp. 116-124, 141; Algar, p. 23; Rizvi 1990, p. 162; Damrel, p. 275). Nevertheless, he as well was in need of a clearly defined image of the enemy, which he found in the Shiites. His attacks against them finally led to his assassination. It can be assumed that his murder was carried out at the order of Najaf Khân, the leader of the political fraction opposing the Rohilla Afghans, who looked up to Mirzâ Maẓhar Jân-e Jahân as their spiritual leader (see Fusfeld, p. 146; Quraishî, pp. 48f, 75ff).

Ghulâm 'Alî and his khânaqah

His main successor *(khalîfah)* was Shâh Ghulâm 'Alî, also known as Shâh 'Abd Ullâh Dehlawî (1743-1824), a Punjabi belonging to the Qadiri order. He also studied at the feet of Shâh 'Abd al-'Azîz, later initiated the construction of a *khânaqah* at the grave of his leader (*murshid*) near Chitlî Qabr (near Sayyid Rifâ'î Ṣâḥib kî Masjid) and carried on the quietistic tradition. With a mosque and facilities for accomodation, this *khânaqah* soon was to become the centre of Mujaddidi ideas and propaganda and in the course of time turned into a port of call for many pilgrims. This was possible because Ghulâm 'Alî transfered his activities and influence to other towns and countries, especially after the annexation (see Fusfeld, p. 74 et passim; Digby, p. 174; Damrel, p. 275; Rizvi 1982, pp. 449ff). It is said that his *khânaqah* offered lodging and food to at least 500 pilgrims (Khân, II, p. 17). The number of Sarâ'îs (inns) bears testimony to a bustling pilgrimage activity. However, due to colonial impact, the number of pilgrims decreased from 100,000 in 1820 to 30,000 in 1900 (Gupta 1986, pp. 257f).

Waqf and other means of support were, however, rejected in conformity with Maẓhar Jân-e Jahân, who had put the rich in the pillory. The economic basis of the *khânaqah*, therefore, were alms and other voluntary donations (*zakât, futûḥ*), funds which often were shared among the neighbors and the visitors (see Fusfeld, passim; Quraishî, p. 91; Tukallî, pp. 322ff; according to Sanderson, p. 51, No. 88, the *khânaqah* was not a *waqf*). The *khânaqah* thus was not only a spiritual centre but also satisfied the basic needs of its visitors. Due to its intentional uncertain economic situation it stood in contrast to the centre of Shâh 'Abd al-'Azîz.

At present it is quite difficult to trace the social basis of the *khânaqah* due to a scarcity of sources on this subject. The Rohilla Afghans (see Fusfeld, passim), who, in 1788, had occupied Delhi for a short while and were politically incapacitated from 1803, probably made up for only a very small part of those paying homage to the sanctuary. One may suppose that a large number of the regular visitors originated from the area near the south-western part of the city, from regions where the majority of artisans and laborers was situated. Those artisans in particular seem to have frequented the sanctuary who were negatively affected by newly established colonial manufacturies and markets, as, e.g., the Muslim weavers (*mu'minîn*), living in their quarter next to the southern city walls (near Borî wâlen); they were regarded as staunch Hanafites (see

Aḥmad, p. 42 and Sanderson, p. 73, no. 144). The fact that in Ghulâm 'Alî's *khânaqah* mats (borâ kâ farsh) and leather cushions replaced beds (see Khân, II, p. 18f), may indicate that those professional groups engaged in the production of these items were also addressed by the Naqshbandi-Mujaddidis. At the religious level, the relationship between the Naqshbandi-Mujaddidis and the weavers was manifested in the lamentations of Ghulâm 'Alî about their deplorable situation and in his prayers for them. On a more profane, external, symbolic level, it was discernible in his wearing a thick coarse cotton cloth produced by the weavers (Quraishi, p. 90; Tukallî, pp. 322f). In fact, traditional norms and affiliations were often articulated through dress (see, e.g., Joshi, p. 258).

Ghulâm 'Alî's successors were Abû Sa'îd (died 1834) and, following his death, his son Aḥmad Sa'îd (born in Rampur in 1802, died in Medina in 1860). Abû Sa'îd also studied jurisprudence at the feet of the sons of Shâh Walî Ullâh. Due to his mystic inclination and his pronounced love for the Prophet, however, he also developed a sympathy for the Chishti order. In this manner he could succeed in addressing further groups of the Delhi population. In 1857 he migrated to Medina (see Fusfeld, pp. 162, 200ff).

Tension between the representatives of the Naqshbandiya

Competition, even rivalry, marked the relations between the representatives of these two branches of the Naqshbandiya, due to differences in the motivation for activities and the public addressed and due to the claim of agencies. The quietistic Mujaddidis pursued a mystic reform more oriented towards the inner self, postulated moral action among their followers and simultaneously engaged in integrating local traditions and norms. Even Mazhar Jân-e Jahân accepted Hindus in certain cases (Friedmann). Moreover, Maẓhar Jân-e Jahân must have appreciated the local culture for economic reasons as well: a certain Hindu called Kewal Râm Baniyah had set up a lodging for the Mirzâ near the Jâmi' Masjid. In spite of his propagated elimination of so-called un-Islamic innovations (*bid'at*), even Shiite followers could be made out in his *silsilah* (see Quraishî, pp. 34, 117). Ghulâm 'Alî as well had certain sympathies for local, Hindu rites (see Rizvi 1982, pp. 552-554).[22] The followers of Shâh 'Abd al-'Azîz, in contrast, openly propagated a purist idea of Islam, soon manifesting itself in the *Jihâd* movement.

This difference in attitudes also had its effect on, or rather was based in, their relationship with the colonial rulers: Shâh Ghulâm 'Alî rigorously refused any contact with the British and thus isolated himself from the politically dominant culture. He therefore also rejected the employment of sympathizers and students of Shâh 'Abd al-

22 This apparent contradiction may, for one, be valued as an expression of competition with the purist branch of the Naqshbandiya (i.e. Shâh 'Abd al-'Azîz). It may also stem from the tensions originating in societal expections on the one hand and individual attitudes on the other.

'Azîz within the colonial administration and jurisdiction. In contrast, after Shâh 'Abd al-'Azîz had regained control over the landed property of his father in 1806, his *fatâwâ* had become more docile; five months prior to his death he was even proposed for the post of principal of the Delhi College (Rizvi 1982, passim; Fusfeld, pp. 165f), formerly Madrasah-e Ghâzî ad-Dîn, established in 1792. These tensions between different representatives of Islam frequently expressed themselves in disputes.

Disputes and Confrontations

Discourses, disputes and dogmatic discussions were common and were usually staged on the occasion of the Friday sermon, the *khuṭbah*, not only among the two branches of the Naqshbandiya. Disputes served the purpose of mobilizing one's own group for common activities. However, not all of them were permitted by the colonial masters. The speeches of purists, which might have provoked an open confrontation between Muslims, were in fact prohibited, e.g., those by Shâh Muḥammad Ismâ'îl.[23]

The successor of Shâh 'Abd al-'Azîz was Muḥammad Isḥâq, who was soon engrossed in polemical disputes and later migrated to Mekka, where he died in 1846. In Delhi he left behind a large number of religiously influential followers (see Aḥmad, pp. 167, 412f; Khân, II, pp. 91f, 271f; Rizvi 1982, pp. 94ff). He severely criticized the prevailing practice of innovation (*bid'at*) among the Muslims in Shâhjahânâbâd. This prompted just as severe apologetic essays by Aḥmad Sa'îd, third successor to Chitlî Qabr, defending the rites of popular belief (see Fusfeld, pp. 230 ff).

These inter-Islamic disputes seem to have increased after British entry. They were, however, hardly ever staged in the centre of the city. Frequently the religious scholars staged them within their own (elite) mosque or within their own *maḥallahs*. Only when confrontation had reached a high level of polemic did the opponents meet in the Jâmi' Masjid. One such dispute involving the entire city was staged in 1824, when the representatives of the traditionalists, the Hanafites, encountered the purists or so-called Wahhabis. The controversy caught fire once again on the issue of popular religious practices and the Prophet-cult (see Khân, II, pp. 79-81, 269; Ṣâbrî, pp. 244f; Rizvi 1982, pp. 485, 517ff). Later, around 1850, there were also discussions between Christian missionaries and Muslims staged at the Jâmi' Masjid (see Gupta 1981, p. 8).

Functions of the religious scholars

These hints may suffice to direct some attention to the inter-Islamic disputes virulent at the beginning of the 19th century in Shâhjahânâbâd. Let us now briefly note the functions of the religious scholars in this complex society before coming to

23 Shâh Ismâ'îl Shahîd, who himself stood in the *manqûlât*-tradition, had vehement disputes with scholars of the *ma'qûlât*-tradition.

concluding remarks. It has been argued that in the course of time the transmitters of culture had succeeded in bringing certain regions within Shâhjahânâbâd under their influence. This did not automatically result in the exclusion of other religious groupings from these regions. There was also lively contact between the groups of artisans and traders crossing the boundaries of the *maḥallah*s. Their economic and social interests and interrelations often, in fact, superseded their religious affiliation, which was symbolically represented by religious dignitaries. Within the city, nevertheless, specific areas of settlement for each Islamic tradition can be identified, each recruiting its members from different social strata and professional groups with their particular interests and specific cultural articulations. The religious scholars usually continued to live in their *maḥallah*s, which were comparatively little frequented by the colonialists, and in isolation from the further local surroundings. Their individual worlds existed exclusively within the bounds of their respective *maḥallah*s and therefore unfolded within the *maḥallah* tradition. This exclusiveness was favored by their practice of hardly ever leaving their place of preaching (*takiyah*, *gaddî*) except for pilgrimage and study tours: They rarely paid visits outside the *maḥallah*, their ritual habits were staged at their *dargâh* (*tawajjuh*) or within their mosque (*ṣalât*). It was their disciples who would seek their presence and project their manifold problems, wishes and hopes onto the dignitary. He was the spiritual centre of the *maḥallah*, with blessings and the holiness (*baraka*) originating from him; here the inhabitants of the *maḥallah* would meet for ever repeated rites and festivities in a centre that provided the *maḥallah* with its essential identity and integrated them into a common social and ritual world. The guardians of religious institutions therefore were powerful and influential men with functions as important if not even more important than those of the *mistrî*s (for the role and function of *mistrî*s, see f.e. Joshi, pp. 252f). By establishing a closed unit with its variety of professional groups, complementing each other, with schools, stores for essential goods, etc., each *maḥallah* gave an incentive for a life in seclusion and devotion to God, thereby forming a moral community. Trans-*maḥallah* movement was thus not necessary in this strictly demarcated cultural framework. These patterns may remind us of similar structures in the *qaṣbah*s. There, a tradition of services with particular products and artisanry had developed, a tradition which was legitimized by religious men and institutions. The *maḥallah* as well may have had some notion of the proud *waṭan*, which was characteristic for the *qaṣbah*. The identity of the quarter was thus marked out by its economic and social contacts as well as religious affiliations and therefore often served as a first port of call for new arrivals to the city: "An inhabitant ... traditionally relates to his Mohalla ... (which) is also a social unit that relates to the occupation, trade, language, religion or geographical origin of the Mohalla dwellers ... Thus a migrant ... had a place to which he could go..." (Fonseca, p. 75). This can also be proven for scholars who went to Delhi to study or for the purpose of visiting certain tombs and shrines or to pay homage to a certain holy man, as is expressed in the account of a journey by Sayyid `Abd al-Ḥayy (died 1923).

Conclusion

From the preceeding arguments, the following conclusions can be drawn about the spatial distribution within Shâhjahânâbâd: sacred institutions provide a guide for localizing this complex society. Mosques of different sizes are laid out according to the hierarchical structure of the city, which is rationally and functionally outlined with "'high ranking' trades close to the core and less esteemed trades and professions in greater distance to it." (Ehlers 1991, p. 89). Almost each professional group displays its own mosque, erected by its members or a member of the nobility in their own residential quarter, and usually bearing the name of the donor, of the professions or of the goods produced and/or sold here. The quarters often bear names of the (dominant) professional groups, but sometimes comprise more than just one profession and consequently, often accomodate several mosques. Therefore, it seems meaningful to replace the denotation "*maḥallah*-mosque" with "profession-mosque".

Religious scholars and dignitaries represent the cultural articulations of a professional group to the external areas (the worldly side) of the *maḥallah*, i.e. the trans-*maḥallah* or trans-*galî*, as well as to the internal areas (the Hereafter), i.e inter-*maḥallah*. They usually draw on the social and economic basis of their local community. Through their scholarship and their spirituality they transmit to the population living in the homogeneous quarters an important meaning and the motivational power to act and thus constitute the moral authority of the local population. Within the *maḥallah*'s surroundings, which are stringently divided into different strata and just as rigorously hierarchized, religious dignitaries have the function of endowing the quarter with an identity and therefore, with their ritual centres—usually *waqf*—, constitute the spiritual focus of the *maḥallah*. Their institutions can be localized in specific regions of the city and vary according to their social bases: In the southeast of the city, along the "primary street" at Faiḍ Bâzâr (near Dukân Natû Mistrî) and Khâṣṣ Bâzâr, there are two large mosques and some shrines as well as some religious schools. At the beginning of the 19th century, most of these religious institutions were dominated by one religious grouping, i.e. the Naqshbandis, established around the rather exoteric—simultaniously sufic—teachings of Shâh `Abd al-`Azîz. The trans-regional economic activities prevailing here, the central location, the proximity to the imperial palace and soon also to the colonial rulers as well as the existence of the Kashmiris hint at bustling communication and missionary activities—in fact, among the Naqshbandis, many missionaries were active in special regions, Kashmir being one of them (see Digby, pp. 197, 203)—and presuppose a specific religious culture among the Islamic dignitaries. As demonstrated above, these were integrationist religious scholars, trying, much in conformity with Naqshbandi tradition, to achieve access to the leading strata of society and, by doing so, to obtain influence with them. Thus, they appeal to certain groups, which they supposedly recruit from the closer surroundings. Although at present data seems to be very scarce on the social basis of the movement of Shâh `Abd al-`Azîz and his successors during their activities in the Akbarâbâdî Masjid, one may deduce that it was the intellectual circles, i.e. representatives of an educational elite—hence, the Maḥallah-e Muftiyân and the region named after the Qâḍîs in close proximity—and "high ranking" professionals engaged in trade and media. Whether those traditional professional groups which were gradually pushed aside by colonialism also became a target of this group cannot be said at this stage.

In contrast to this, the centre of the quietistic Naqshbandiya-Mujaddidiya is located in the commercial and cultural core of the city, at the point of interjection known as Bâzâr Chitlî Qabr. At the beginning of the 19th century, this region had been largely spared from colonial influence owing to the difficulty of access. The Bâzâr Chitlî Qabr had developed around the monumental graves of saints and up to this day this area can hardly be explored by strangers: narrow *galîs* and corners determine the picture of this multicoloured medley, limited however to local trade and artisanry. The local catchment area of the Naqshbandiya-Mujaddidiya spread from Chitlî Qabr via Dargâh Shâh Turkmân, the axis Sîtârâm Bâzâr up to Ajmeri Gate in the southwest and up to Turkman Gate and Dehli Gate in the southeast, an area which in some parts has the highest number of professional groups, a fact which can also be derived from the large number of mosques in the area.

Besides these two Islamic traditions, there naturally existed many others, e.g., the Shiites to which we have referred on several occasions. As can be seen from the map, they were not limited to just one area in the city but rather spread about town, except—strangely but according to the map—in areas accommodating artisans. Even if they had some religious institutions[24] to practice their rites, the Shiite community strangely does not seem to have had a common ritual centre. The erratic dispersion does not, however, necessarily point to an unproblematic integration of this sect into the society of Delhi. In fact, there were repeated vehement quarrels between the Sunnis and the Shiites, reaching a climax with the polemical essay "Tuḥfah-e Ithnâ-e `Ashariya" by Shâh `Abd al-`Azîz, which was declared compulsory reading among the Naqshbandis and which was equally responded to by the Shia Ḥakîm Mirzâ Muḥammad Kâmil. In some cases quarrels were even of violent nature, usually and traditionally starting at the time of the Muharram processions. Mirzâ Maẓhar Jân-e Jahân fell victim to one such clash. Moreover, the Shia did have a cultural centre in Lucknow, competing with Delhi culturally, politically and economically. Therefore, one might have expected that the Shia would have settled in just one specific area, similar to other religious groups, in order to form a single and powerful group. The reason for its irregular dispersion over the city therefore must be sought in other suppositions than the one of its being an integrated part of the culture in Delhi. Their spreading in the entire core space of the city might have been possible due to the practice of *taqîya*, a practice which

---

24 E.g., mosques southeast of Chândnî Chawk (near Nawwâb Ṣâḥib kî Masjid and Khûnbahâ kî Masjid) and a few scattered *dargâhs* (near Ṣâbir Bakhsh Garden; near Dargâh Shâh Turkmân; near Biŕ wâlî Masjid) and numerous *imâmbâŕahs*.

again and again outraged Sunni scholars. In this way at least some of the Shia had a chance to achieve common social privileges as a pretended Sunni or even as a member of some other belief. As a Muslim minority they may also have sought the proximity of the court in order to get imperial support.

Representatives of other religious groups were concentrated in still other regions. This was also true after 1857 in the case of the Ahl-e Hadith. They were mainly concentrated in Khârî Bâ'olî, at Ghî kâ Katrah (butter-oil market) and at Phâṭak Ḥabash Khân (gate of Ḥabash Khân), around the religious scholar Nadhîr Ḥusain (died 1902). This is also where the Fataḥpûrî Masjid was located, bordering upon an important quarter of commercial activity, the Ashrafî Bâzâr, as well as upon the financial quarter of the city with the mosque of the Muḥtasib located nearby. The spatial surroundings of this group related to its social origin, mainly as traders and entrepreneurs. For a long period, the Fataḥpûrî Masjid remained the centre of their theological and political activities.[25]

Thus each religious group, according to its professional and social affiliation, occupies specific areas, within the hierarchically arranged city, symbolized by a specific sacred institution. This is why Shâhjahânâbâd, material expression of imperial rule, displays different "Islams" reflecting manifold economic, social and political interests and claims. Up to this very day, this type of regionalization of Islam can be made out at the inner-urban micro-level as well as at the trans-regional macro-level. What is Islamic about this kind of dissemination and spatial pattern except for its Muslim actors, may, however, be an open question.

25 This socio-economic tradition among the Ahl-e Hadith is effective to the present day: even in Pakistan they recruit their followers from areas which are important for economic development, "commercial areas", so-to-speak.

Bibliography

Abu al-Fazl (1891): *A'in-e Akbari*, Vol. II, Calcutta: Asiatic Society of Bengal (transl. by H.S. Jarrett)

Aḥmad, Bas̲h̲îr ad-Dîn (1919): *Wâqi'ât dâr al-Ḥukûmat Dehlî*, Vols. I-III, Agrah: S̲h̲amsî Press; here only vol. II

Alam, Muzaffar (1986): *The Crisis of Empire in Mughal North India*,Delhi: OUP

Algar, H. (1990): "A Brief History of the Naqshbandî Order", in: Gaborieau et al., pp. 3-44

Bayly, C.A. (1980): "The Small Town and Islamic Gentry in North India: the Case of Kara", in: K. Ballhatchet/J. Harrison (eds.): *The City in South Asia: Pre-Modern and Modern*, London: Curzon Press, pp. 20-48

Blake, S.P. (1986): "Cityscape of an Imperial Capital: Shahjahanabad in 1739", in: Frykenberg (ed.), pp. 152-191

Chandhoke, S.K. (1991): "The Delhis within Delhi", in: *TRIALOG*, Vol. 29, pp. 13-16

Chandra, S. (1986): "Cultural and Political Role of Delhi, 1675-1725", in: Frykenberg (ed.)

Damrel, D.W. (1990): "The Spread of Naqshbandi Political Thought in the Islamic World", in: Gaborieau et al., pp. 269-287

Digby, S. (1990): "The Naqshbandîs in the Deccan", in: Gaborieau et al., pp. 167-207

Ehlers, E. (1978): "Rentenkapitalismus und Stadtentwicklung im islamischen Orient. Beispiel: Iran", in: *Erdkunde*, Vol. 32, pp. 124-142

Ehlers, E. (1991): "The City of the Islamic Middle East", *Colloquium Geographicum*, Vol. 22, pp. 89-107

Fonseca, R. (1971): "The walled city of Old Delhi", in: *Ekistics* No. 182, pp. 72-80

Frykenberg, R.E. (ed.) (1986): *Delhi Through the Ages: Essays in Urban History, Culture and Society*, Delhi: OUP

Friedmann, Y. (1975): "Muslim Views of Indian religions", in: *Journal of the American Oriental Society*, 95/2, pp. 214-221

Fusfeld, W.E. (1981): *The Shaping of Sufi Leadership in Delhi: The Naqshbandiyya Mujaddidiyya, 1750 to 1920*, unpubl. PhD. thesis, Univ. of Pennsylvania

Gaborieau, M./Popovic, A./Zarcone, T. (eds.) (1990): *Naqshbandis; Historical Developments and Present Situation of a Muslim Mystical Order*, Istanbul: Institut Francais d'Études Anatoliennes

Gupta, N. (1981): *Delhi between two Empires, 1803-1931: Society, Government and Urban Growth*, New Delhi: OUP

Gupta, N. (1986): "Delhi and its Hinterland: the Nineteenth and Early Twentieth Centuries", in: Frykenberg (ed.)

Ḥabîb Ullâh (1988): *D̲h̲ikr-e Jamî'-e Awliyâ-e Dehlî*, ed. by S̲h̲arîf Ḥusain Qâsimî, Tonk: Arabic and Persian Research Institute

Ḥayy, Sayyid 'Abd al- (1979): *Dellî awr us ke aṭrâf*, ed. by Ṣâdiqah D̲h̲akkî, Lâhawr: Maktabah K̲h̲alîl (repr.)

Joshi, C. (1985): "Bonds of community, ties of religion: Kanpur textile workers in the early twentieth century", in: *The Indian Economic and Social History Review*, Vol. 22/3

K̲h̲ân, S. Aḥmad (1990): *At̲h̲âr aṣ-Ṣanâdîd*, Vols. I-III, ed. by K̲h̲alîq Anjum, Dillî: Urdû Akâdemî (repr.)

Malik, J.: "Gelehrtenkultur in Nordindien", in: *PERIPLUS 1992*, Münster: LIT Verlag, pp. 152-163

Malik, S.J. (1990): "*Waqf* in Pakistan; Change in Traditional Institutions", in: *Die Welt des Islam*, Leiden: Brill Vol. 30, pp. 63-97

Qâdirî, Muḥammad Ayyûb (1982): *Hindûstân men Muslim Firqah wârîyat*, Karâchî: Idârah-e Ma'ârif al-Ḥaqq

Qâsimî, 'Attâ ar-Raḥmân (1989): *al-wâh aṣ-Ṣanâdîd*, Dehlî: ShâhWalî Ullâh Akâdemî

Quraishî, 'Abd ar-Razzâq (1979): *Mirzâ Maẓhar Jân-e Jahân awr un kâ Kalâm*, A'ẓamgarh: maṭba'-e ma'ârif

Richard, J.F. (1986): *Document Forms for Official Orders of Appointment in the Mughal Empire: Translation, Notes and Text*, E.J.W. Gibb Memorial Series, New Series, XXIX, Cambridge: The Burlington Press

Rizvi, S.A.A. (1982): *Shâh 'Abd al 'Azîz: Puritanism, Sectarian Polemics, and Jihad*, New Delhi: Munshiram Manoharlal

Rizvi, S.A.A. (1983): "Shâh `Abdu'l-`Azîz's Madad-i Ma`âsh in Delhi and the British", in: M. Isreal/N.K. Wagle (eds.): *Islamic Society and Culture: Essays in Honour of Professor Aziz Ahmad*, New Delhi: Manohar Publications, pp. 135-147

Rizvi, S.A.A. (1990): "Sixteenth Century Naqshbandiyya Leadership in India", in: Gaborieau et al., pp. 153-165

Sanderson, G. (1916): *List of Muhammadan and Hindu Monuments, Vol. I Shahjahanabad*, Calcutta: Superintendent Government Printing

Ṣâbrî, Imdâd (1972): *Dehlî kî yâdgâr hastiyân*, Delhî: Jamâl Printing Press

Siddiqi, Z. (1963): "The Muhtasib Under Aurangzeb", in: *Medieval India Quarterly*, No. 5, pp. 113-119

Trimingham, J.S. (1971): *The Sufi Orders in Islam*, Oxford: OUP 1971

Troll, C.W. (ed.) (1989): *Muslim Shrines in India; Their Character, History and Significance*, Delhi: OUP; see review by J. Malik in: *Die Welt des Islam*, Vol. 32/1992, pp. 161-164

Tukallî, Muḥammad Bakhsh (1976): *Tadhkirah-e mashâ'ikh-e naqshbandîyah*, Lâhawr: Ma'ṣûm Akîdîmî (repr.)

Contemporary Old Delhi: Transformation of an Historical Place

Thomas Krafft (Marburg)

After the demise of the Mughal Empire and the quelling of the uprising of 1857, the third decisive hiatus in the development of Shâhjahânâbâd/Old Delhi was Independence and the subsequent partition of the subcontinent.

The bloody interchanges between the religious groups, the mass exodus of large portions of the Muslim population, the still greater influx of refugees from Punjab and the rapid growth of Delhi in the following period caused a radical break with the existing structures. In many quarters of the Walled City an almost complete exchange of population took place. In this respect and with reference to their political and economical influence, we must speak of a marginalisation of the Muslim portion of the population. There is no doubt therefore that since 1947, at the latest, Islam in the form of a religious culture and law had little or no influence on the city development. In spite of this, in formal and functional structures, Old Delhi still shows great similarities today, with the traditional city centres of comparable cities of the Islamic Orient. Certain areas are, as in former times, shaped by the large Muslim part of the population: walking from Urdu Bazar (southern side of the Jama Masjid) along the Matia Mahal and the Shah Abdul Khair Marg in the direction of the Turkman Gate or turning into the system of lanes and alleys, one has the feeling one has been transported to the Islamic Orient, from the atmosphere of mosques, madrasas, bazaars, language, clothing and sounds.

The mahallah system and Islamic tradition have been preserved in this area because the defensive Muslim minority clung to symbolic ties and rules. From the very beginning a considerable number of Hindus lived within the Walled City. In fact: till the mid-nineteenth century both communities (Hindus and Muslims) were of about the same size (cf. Gupta 1981). The intrusion of large numbers of Hindus into former Muslim quarters after Partition led to severe tensions, resulting in frequent riots. Nevertheless, most Muslims view the Walled City as a safe place which they identify with the symbols of Islamic culture. And in times of rising communal conflicts Muslims living in other places show an increased readiness to move to the Walled City, although most communal conflicts in Delhi originate in this area.

In the following the focus is on the causes of physical decay and functional change which have led to a radical transformation of Shâhjahânâbâd/Old Delhi.

Population

During the census decade 1941-51 the population of Old Delhi more than doubled. The following decade is also marked by a further, although significantly more moderate, increase in population. Since the mid 1960s the trend has reversed and population figures for the Walled City have decreased[1] slightly (Fig. 1). Three partially overlapping and partially opposing processes mark the development as a whole:

- a complete upheavel of the population structure as a result of the partition of the subcontinent,
- congestion and housing shortage creating slum-like conditions in many quarters,
- and finally commercialisation and crowding out of residents.

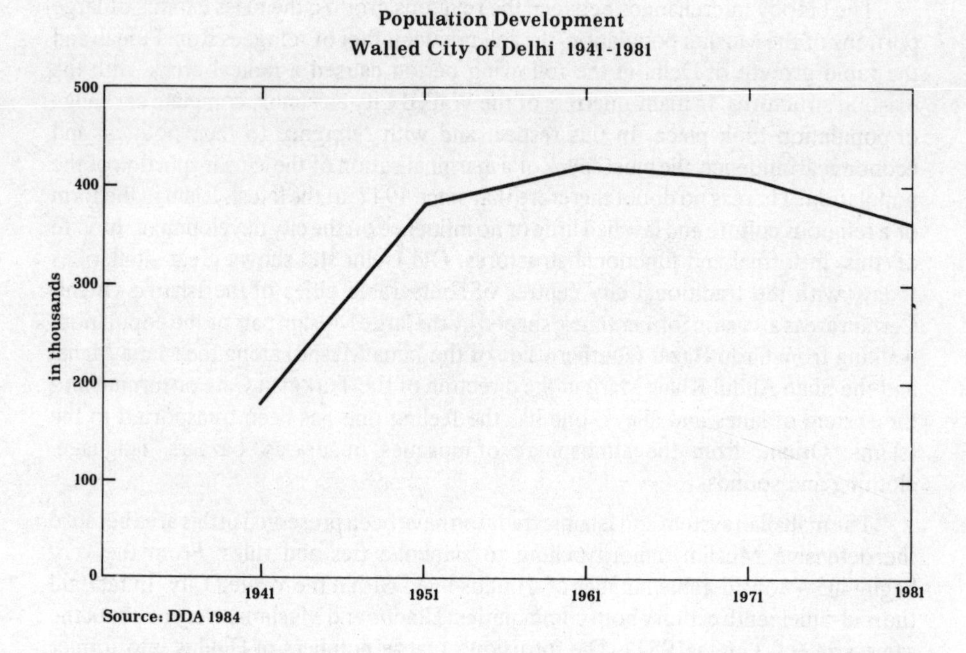

Fig. 1: Population in Old Delhi

1 The data on population in Old Delhi must be viewed with distinct reservations. Because the foremost goal of planning in this district was to reduce the residental population, the figures take on a special significance for assessing the success or failure of urban planning institutions. While viewing the data published by these institutions this context should be considered. The figures at issue should therefore be understood as references to general trends rather than as absolute values.

Partition and the bloody unrests between 1947 und 1948 created tremendous changes in Delhi's population structure. 329,000 Muslims left the Delhi region between 1947 and 1951. In return, during the same period 495,000 refugees from Punjab came to Greater Delhi. The opposing refugee streams alone resulted in a population increase of 166,000. In addition to this, during the same short period of only four years "normal" migration and natural increase added another 206,000 people to Delhi's population. (Rao 1965, p.56)

Old Delhi was especially hit by this development. Traditional quarters which had been shaped by their Muslim occupants over the centuries, as in the district of Mori Gate/Kashmere Gate, lost the greater part of their occupants in a very short time. Houses and flats rarely stayed empty for long, however. They were soon taken over by refugees—partially legally through allocation, but mostly illegally.

The effects of these changes on Old Delhi can be summarised as follows:

Fundamental Changes in the Ownership Structure

Thousands of houses and pieces of property changed ownership in the first years after Partition. Illegal takeovers and allocation of so-called "evacuee properties" through the Ministry of Relief and Rehabilitation offered the quickest solutions to the acute housing crisis. The less needy also took advantage of this easy opportunity to gain their "own" property. Tenants took over the houses of landlords who had fled; neighbours claimed adjacent plots and dealers the businesses of competitors who had fled. The numerous nontransferable wakf properties in formerly Muslim quarters found new "owners". As a result, even today there is confusion as to the actual ownership of numerous buildings and pieces of property.

By transferring hundreds of evacuee properties to the urban planning authorities the Ministry of Relief and Rehabilitation created the requirements for some slum redevelopment measures or rather the building of public amenities, but still today—more than 40 years after Partition—a large number of property ownership disputes are lodged with the responsible courts. Several hundred of these were instituted by the Delhi Wakf Board alone.

Marginalisation of the Muslim Population

For the Muslims remaining in India the exodus of a large portion of the community to Pakistan had decisive consequences. Almost the entire upper political strata left India hoping for better opportunities for their future life in Pakistan, as proclaimed by the protagonists of the Two Nation Theory. Since much of the Muslim middle class also left for Pakistan, it was mainly the Muslim lower class that remained in India.

Today India's Muslims constitute with about 12% of the total population the largest minority group. And in spite of their regional, linguistic, cultural, social and sectarian heterogeneity they do have the characteristics of a cohesive religious minority group. As a whole the Muslim community is an economically and educationaly backward section of Indian society, lacking adequate political representation and influence. For most Muslims the acknowledged backwardness[2] is—at least in parts—the result of Partition and the following communal prejudices (Ansari 1989).

The trend towards political and socio-economic marginalisation of Indias's Muslim population after 1947 is represented in Delhi as well. Most of those who stayed on in the city belonged to the economically weaker section of society, had little education and were bound to traditional customs and values[3] becoming easy fodder for frequent communal riots.

Beginning Commercial Transformation

While Delhi Muslims saw themselves confronted with political and social degradation as a result of the partition and reacted with regression and lethargy, it was comparably easy for Punjab refugees to consolidate their position in Delhi, become established and compete effectively with the Delhiwallas.

Many of the refugees originated from the urban middle class and were well trained and educated.[4] Thus illiteracy ranged several per cent lower among the refugees than among the indigenous Delhi population. Many Punjabis seized the chance to build up a livelihood in trading almost from nothing within a fairly short time. And the centre of this process was the Walled City. Within a few years Punjabi traders came to be a major force in the local economy of C̲h̲ândnî C̲h̲awk, Kashmere Gate and New Lajpat Rai Market. (Datta 1986)

In 1958 74,5 % of the enterprises operating in Old Delhi were founded after Independence and Partition. (Aziz 1983, p.61). Buildings immediately available in Old Delhi were occupied and used for all kinds of business regardless of their original purpose, merely according to the respective needs. Thus the basis for the contemporary problems in urban land management was already laid in the 1950s with this first phase of unplanned and uncontrolled commercial growth.

---

2 For a more differentiated view of the extent and cause of backwardness see the various articles in Ansari (1989).

3 Under direction of the Government the Census of India ceased to publish detailed data relating to religion. The actual size, spatial distribution, and economic status of the different religious communities in Delhi can only be estimated.
According to recent estimates Muslims constitute about one third of Old Delhi's population and about 8% of Delhi's total population.

4 According to Rao (1965, p112 f.) the ratio of urban to rural refugees was 4:1.

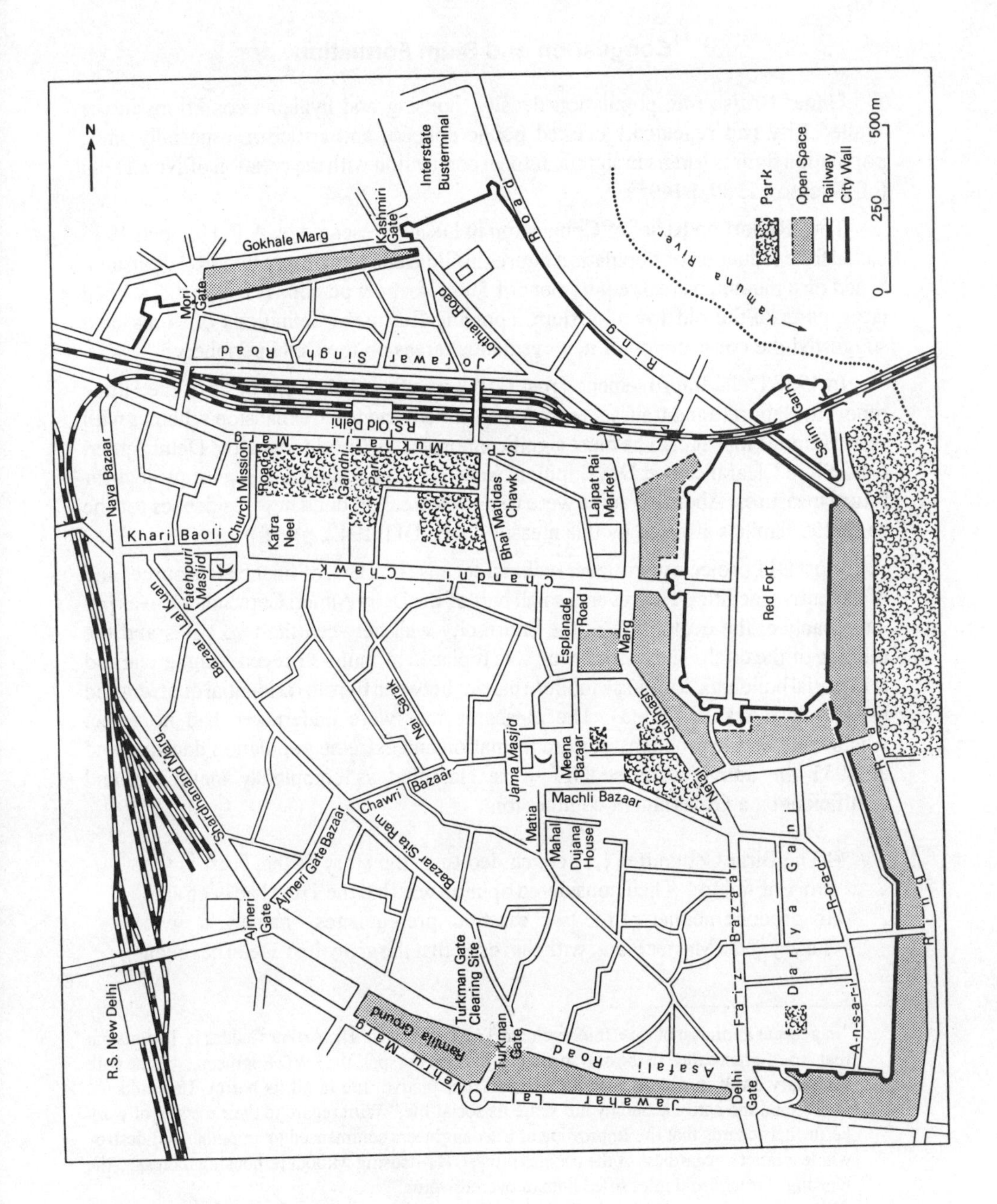

Fig. 2: Contemporary Old Delhi

Congestion and Slum Formation

Under British rule population density, housing and hygienic conditions in the Walled City had repeatedly evoked public concern and criticism, especially since population figures had risen significantly in connection with the creation of New Delhi (cf. Evenson 1989, p149f.).

The "Report on Relief of Congestion in Delhi" presented by A.P. Hume in 1936 called for a reduction of population figures in Old Delhi by at least a hundred thousand, based on a minimum areal requirement of 50 sq. feet per person. The report classified large parts of the old town as slum, applying European definitions of slums, and suggested the construction of new residential areas on the Delhi periphery.[5]

In 1937 Delhi Improvement Trust (DIT) was founded on the basis of the Hume report to elaborate and transact town improvement and town expansion schemes with A.P. Hume as manager. The most significant project in the old town, the Delhi Ajmeri Gate Slum Clearance and Development Scheme, was designed as a pilot project for future measures. About 70 acres were to be "cleared" to build new residences for the ca. 2,400 families affected by this measure. (see DIT 1942, p.17f.)

Most DIT projects were never realized because of a lack of financial resources and of executive facilities. However, a small part of the Delhi Ajmeri Gate Scheme was put into practice: the demolition of the entire city wall between the two gates and the refilling of the ditch. The historic wall was replaced by multi-storeyed commercial and residential buildings, forming a modern barrier between the old residential quarters and New Delhi (Asaf Ali Road). The measures that were undertaken had no effect whatsoever on the problems of slum formation and extreme population density. And in 1951 the activities of the DIT were classified as completely inefficient and insufficient by a Government commission[6]:

> "The Birla Committee [...] concluded that 'the story of the Trust is the story of failure'. Their considered opinion was that the Trust had been able to produce neither of the two essential prerequisites—namely, a civic survey and a Master Plan, 'with the result that the growth of Delhi has been

---

5 "in a recent appraisal of this 'Improvement Trust Approach' Dr. Arthur Geddes [...] points out that "each Trust called in the only available 'man on the spot' the PWD Engineer ... but he with his salary dwelt in distant civil lines, aloof from 'native' life in all its horror. He could not understand the native economy nor share its social life." With regard to their method of work he further records that the Improvement trust engineers commenced to amputate and destroy whole quarters, regardless of the peoples protest. De-housing without re-housing increased the housing shortage and intensified human overcrowding."
(Bharat Sevak Samaj 1958, p.218)

6 The so-called Birla Committee headed by G.D. Birla. See: Report of the Delhi Improvement Trust Enquiry Committee, New Delhi: Government of India Press 1951.

proceeding in a haphazard way, with little foresight and imagination and without any coordination'."
(Bharat Sevak Samaj 1958, p.218)

The 1950s saw a drastic aggravation of urban problems in Old Delhi due to the massive population growth after 1947 and to the constant rise in demand for commercial space. A report on "Slums of Old Delhi" presented in 1958 (Bharat Sevak Samaj) estimates a slum population of 50,000 in katras[7] and of 12,000 in four shanty areas (Bastis) within Old Delhi.

Slum-like living conditions were to be found in other residential areas as well. Urban development authorities used age and condition of buildings, infrastructure, hygienic conditions and, most important, population density (in persons per acre) as criteria for the classification of "slums". A.P. Hume and the DIT had set 200 ppa as the maximum admissible figure. The limit was raised to 250 ppa in almost every part of Old Delhi in the Delhi Master Plan, which was enforced in 1962 to work as a planning guide for the following two decades (up to 1981). The actual population density exceeded the limit at the time of enforcement in all but one residential planning zone (A-20 Darya Ganj[8]).

The designation of slum areas was carried out on the basis of the Slum Areas (Improvement and Clearance) Act of 1956. In 1974 this task was handed over from the Municipal Corporation of Delhi (MCD) to the Delhi Development Authority (DDA), which has divided the slum areas into the following three categories:

1. *Conservation Area.* The conservation area is the area which is basically fit for human habitation, and with minimum repairs, the area can be kept conserved and preserved for the coming years.
2. *Rehabilitation Area.* This is the area which requires some redevelopment schemes in order to achieve the minimum living conditions. These are the areas where substantial financial investment is needed for improvement.
3. *Clearance Area.* This is the area which is totally unfit for human habitition and needs to be pulled down for rebuilding. In this area, total demolition is needed

7 "Katras [...] are extremely unhygienic structures with no proper ventilation, no drainage, no latrines and are in their existing state unfit for human habitation. It may be observed, however, that most of these Katras are single storeyed and a number of them have open spaces within them, which implies that while there may be overcrowding in the built up area (living space) the density per acre should not prove unmanagable."
(Bharat Sevak Samaj 1958,p.216)

8 As a result of the destruction of the traditional urban fabric by the British during the 19th century the building structure in Darya Ganj is not as old and is in better condition than in most of the other parts of the Walled City. The concentration in this area of several medical and welfare institutions, secondary schools, and other educational facilities, most of them based on religious endowments or private foundations, hindered the progress of commercial encroachment.

and an entirely new colony with proper sanitation and civic facilities is to be developed by the Authority. (Pillai 1991, p.80)

All parts of the Walled City were classified according to the criteria mentioned above within the scope of the Delhi Master Plan. The entire area between the Jama Masjid and Turkman Gate was classified as Clearance or Redevelopment Area. In effect the whole of Old Delhi was declared to be slums, with only slight differences concerning the measures to be taken. This low-profile approach proved to be completely inadequate as a way of solving urban development problems. Eventually, in the new urban development plan "Master Plan for Delhi Perspective 2001" it had to be given up completely (DDA 1987).

The development of the population density between 1951 and 1981[9] shows that none of the decongestion measures were successful. On the contrary, the population density in all Old Delhi planning zones is still extremely high, despite marked differences. The figures for 1981 range between 286 ppa in Chândnî Chawk/Lajpat Rai Market and 807 ppa in Lâl Darwâzâ (near Jama Masjid). Hence the planning target of 250 ppa as a maximum for 1981 was not achieved in any district at all. In eight of sixteen planning zones this figure was more than doubled. In spite of the assumed decrease in total population figures, population density increased in comparison to 1951 and 1971, because residential space was reduced by demolition or commercial encroachment at a much faster pace than population decreased due to out-migration (Fig. 3).

The many pavement dwellers are not considered in these figures. They are to be found mainly in Chândnî Chawk, in Old Delhi Railway Station and in the areas around Jama Masjid. The 1971 census registered 15,000 homeless people in Delhi Municipal Corporation. 10,000 of these were registered in Old Delhi itself. This means an increase in the homeless population 191 % of between 1961 and 1971. Social workers estimate that there were about 15,000 to 20,000 homeless people in Old Delhi in 1990, only a few thousands of whom found room in various night shelters in the Walled City.

If we view population density as an isolated factor, we have to concede that the problem of slum formation has grown continually worse during the last four decades. Slum clearance measures even aggravated the situation, since only about 10 % of the cleared areas were used for the reconstruction of residential buildings. Thus, the shortage of housing space and consequently population density increased even further. (cf. Jain 1990)

9 These figures were published by the DDA Perspective Planning Wing. They represent the "Gross Residential Density (ppa)" derived from the number of inhabitants per planning zone and randomly chosen "Proposed Gross Residential Areas". The actual areas being used for residential purposes are significantly smaller, thus the population density is markedly higher. In 1971 the gross residential area amounted to 707.75 acres while only about 500 acres were actually being used for residential purposes (see TCPO 1974, p.21f).

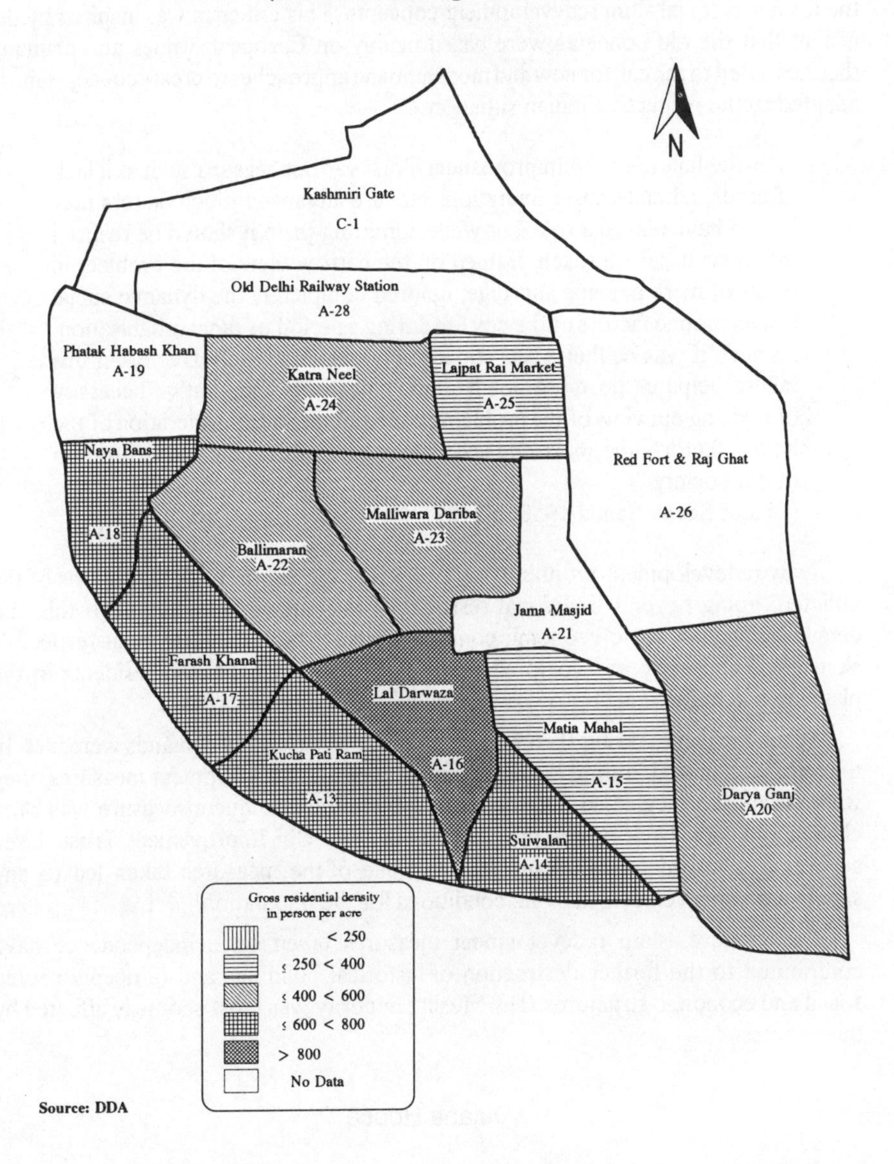

Fig. 3: Population Density in Old Delhi

Slum Redevelopment

In the early years after Independence Indian planners and officials sharply criticised the former colonial slum redevelopment concepts. This criticism was inspired by the insight that the old concepts were based mainly on European values and planning theories. It led to the call for new and more humane approaches to create concepts more adapted to the respective Indian situation:

> "For the failure of the Improvement Trust various reasons such as a lack of funds, administrative limitations etc. are advanced. Such factors may indeed have played a role, but while admitting them it should be realised that their basic approach, framed on the narrow view of the problem in terms of mere housing shortage, ignored completely the dynamic social and economic factors of the city life during a period of rapid urbanisation. As such it was neither a realistic nor a human approach. Deploring this failure helps us no more but it does indicate the imperative necessity of revising our view of the problem so as to admit due consideration of the human factor—the social and economic needs of this, the lowest stratum of our society."
> (Bharat Sevak Samaj 1958, p.218)

New redevelopment programs were to affect the entire living conditions of the afflicted, going beyond traditional restrictions to housing conditions. To this end detailed surveys of socio-economic conditions were to be undertaken in the respective slum areas and early and comprehensive participation of afflicted residents in the planning and enforcement process was called for (ibidem p.219f).

When the Delhi Master Plan was set up only parts of these demands were met. In the following period, during the enforcement of slum redevelopment measures, they were completely neglected. On the contrary, the most frequent measure was slum clearance, falling back on the programmes of the Delhi Improvement Trust. Even according to estimates of urban planners none of the measures taken led to any significant improvements in living conditions for the population.

Furthermore, slum redevelopment measures taken since independence have contributed to the further destruction of historical buildings and of deeply rooted social and economic structures. The Muslim minority was most seriously affected by this process.

Dujana House

Dujana House Rehabilitation Scheme (Bazar Matia Mahal) was among the first redevelopment projects undertaken after Independence. The city palace (nawwâb

Azîzâbâdî-kî-hawêlî) was part of the Moghul property confiscated by the British after 1857 and was renamed Dujana (Doujana) House after its subsequent owner (see Sanderson 1916, p.47).

Some of the buildings were so delapidated that the site was classified as slum and "cleared" in the early 1960s. Residents were put up in "transit camps". About 120 flats were to be constructed in four multi-storeyed tenement buildings. These blocks form an alien architectural element in a formerly homogeneous quarter. They are adapted neither to the climatic conditions nor to the way of life of the predominantly Muslim residents. The project has been only partially implemented, and in retrospect this project was therefore classified as an absolute failure as far as the original planning goals are concerned:

> "Now after a lapse of some twelve years more squatters have occupied the spaces which were either to house other evicted families or to provide the much needed open space. The area is now full of squalor and filth and many unauthorized workshops have since been established. The problem as it is now is more complicated than when this project was initiated."
> (TCPO 1974, p.14)

Meena (Mînâ) Bazar

Up to the mid 1970s there was a bazaar on the open space around Jama Masjid. More than 500 little stalls, some of them directly adjacent to the mosque, made up one of the most interesting and characteristic markets in the Walled City (Mohan 1973). Presumably its development dates back to the second half of the nineteenth century when the bazaar and the residential quarters between Jama Masjid and the Red Fort were destroyed by the British[10]. This bazaar—in combination with the Friday mosque—was one of the cultural and social centres for the Muslim population in Old Delhi. Over 90 % of the traders were Muslim. The entire rent proceeds (about 100,000 Rs. per annum) went to the Jama Masjid, for the stalls were located on the open space belonging to the mosque (waqf).

The Delhi Master Plan—under the title of "beautification"—projected extensive clearance not only of the bazaar area, but of all traditional building structures around the mosque as well. According to this "Jama Masjid Community Square Plan" modern six-storey (!) tenant and commercial buildings were to be constructed along three sides of the square with the Jama Masjid in its centre. The fourth side of the square (with a view of the Red Fort) was projected as an open space (see Master Plan App. J, p.147ff).

---

10 "... after the conquest of Delhi by the British in 1857, the vast shopping complex which had existed from the Red Fort to the Delhi Gate was destroyed by them, to be resurrected once again as an open market around the steps of the Jama Masjid." (Mohan 1973, p.23)

Because the overall plan could not be enforced, urban planning authorities tried to raise acceptance among the population by splitting the project up into isolated measures. This way the number of people afflicted was reduced for each single measure. In November 1975 the entire market area around Jama Masjid was "cleared" by bulldozers after a forced eviction ordered by the DDA. Six months later—under the pressure of massive public protests—construction of a closed bazaar complex (Meena Bazar) with about 370 small shops was started east of the mosque. As the projections were too small from the beginning, the old problems reappeared soon afterwards:

> "Today's Meena Bazar is uglier than what it was in 1976. Not only that, the surroundings of the mosque which were forcibly vacated in 1975 remain full of squalor. [...] Within the Meena Bazar market [...] an equal number of unauthorized shops [...] have come up during the last few years. In addition there are scores of unauthorized pavement hawkers squatted on the narrow entrances to the shops"
> (The Statesman Oct. 18 1988, p.5).

Turkman Gate

The residential quarter near Turkman Gate in the southern part of the Walled City is still seen as the symbol of inhumane and ruthless urban planning policies during the Emergency of the 1970s. This quarter is part of the area that was classified as redevelopment zone by the Delhi Improvement Trust in connection with the Delhi Ajmeri Gate Scheme in the late 1930s. Some of the land was already acquired by the Municipality at that time. The Slum Areas (Improvement and Clearance) Act extended municipal property in the 1950s. The redevelopment plan was sanctioned with some modifications for the second time in the mid 1960s, but its enforcement was suspended for the time being.

In the early 1970s plans were developed to connect Delhi's two main business centres, the modern CBD around Connaught Circus and the traditional bazaars of the Walled City. For this purpose all existing buildings between Jama Masjid and Turkman Gate—the core of the Muslim residential quarters in the Walled City—were to be demolished and replaced by modern multi-storey commercial buildings. This project (which promised enormous speculative profits) could only be started under the specific conditions of the Emergency.

In retrospect it is impossible to reconstruct the originally intended extent and nature of the demolition measures, which were started in April 1976 at the order of the DDA under massive police presence. The available accounts differ greatly and the later statements of the officials involved in the decision making process are too contradictory. The fact is that: 150 houses were demolished in an area of about 6.7 acres, rendering more than 700 families homeless. The evicted were promised and some even provided with housing in remote resettlement colonies. The overwhelming majority of the former residents of Turkman Gate, however, slept in the open close to the clearance site. Demolition had to be stopped after a few days because of fierce resistance from the population. The protests, though powerless at first, escalated into severe riots. In their course several demonstrators were killed by the police, with numerous more injured on April 19. A fairly liberal newspaper summarized the motives of the decision-makers on the occasion of the anniversary of the event as follows:

> "Arbitrary change of land use from residential to commercial purposes was one major reason for these brutalities. The other factor was to disperse the Muslim community to different areas and thus destroy their coherent character which had come up since the 1946-47 communal carnages."
> (The Statesman April 19 1986, p.5)

The second aspect, that is the assumed deliberate discrimination of a religious minority, adds to the explosiveness of this conflict. No matter whether this was indeed a conspiracy[11] against the Muslim community, an extreme case of unscrupulous land speculation or merely a ruthless attempt to carry out a long suspended project under the shield of the state of emergency: as a result of the tense political situation in spring 1976, the conspiracy theory appealed not only to Muslims as a perfectly credible explanation.

Although the authorities gave up the project of building a commercial complex after the Congress Party lost the elections, it took several years until new residential buildings were constructed on this site. And as in the case of Dujana House the new residential blocks form an alien architectural element in the historic setting of Turkman Gate.

Commercialisation of traditional neighbourhoods

Delhi shows the typical dual character of many oriental cities with the juxtaposition of a traditional medina (Old Delhi) and a modern "westernized" city centre (Connaught Place/New Delhi). Government and administration, office functions and financial services, and most of the modern retail trade are concentrated in New Delhi. However, the Walled City has not lost its economic importance and function. On the contrary, in the last decades Old Delhi has experienced a dramatic increase in commercial activities, especially in wholesale trade . Delhi is one of the leading trade centres of the subcontinent and serves as the major commercial centre, both wholesale and retail, for all of northern India. More than 50 % of Delhi's wholesale enterprises are located in the congested central bazaars of Old Delhi (Fig. 4).

---

11 In particular the (alleged) personal involvement of Indira Gandhi's son and designated successor, Sanjay, gave rise to numerous speculations.

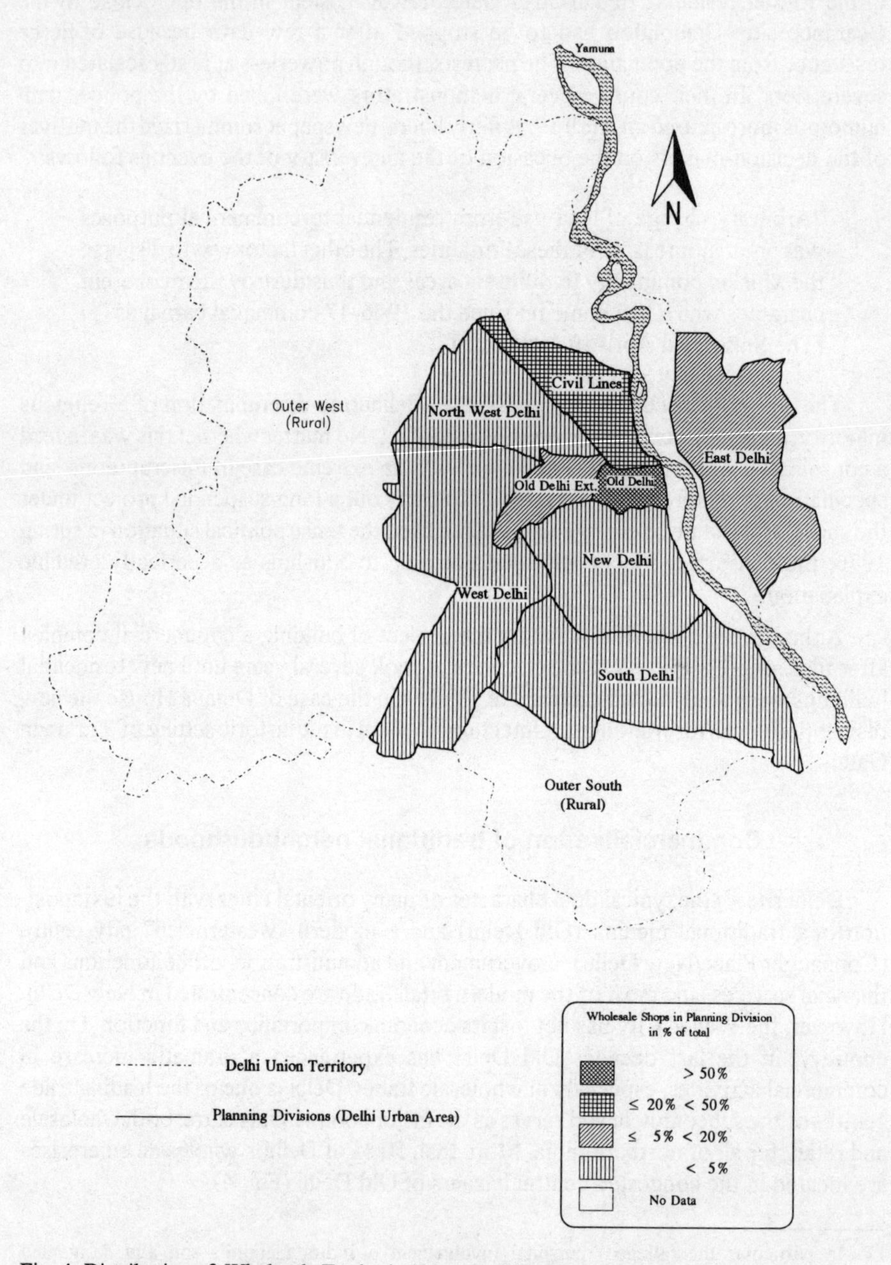

Fig. 4: Distribution of Wholesale Trades in Urban Delhi (1981)

Wholesale trade in the Walled City encompasses more than 20 commodity groups. Amoung these the most significant are the markets for grain (Nayâ Bâzâr - Khârî Bâolî), construction materials (Ajmeri Gate), iron and hardware (Châwrî Bâzâr - Lâl Kuân), electronic goods (Lajpat Rai Market), paper (Châwrî Bâzâr), dry fruits and spices (Khârî Bâolî), and vehicle and motor parts (Kashmere Gate). Less than 25 % of the transactions conducted in these commodities serve the National Capital Region.

Of overwhelming significance is the trade in textiles, which is mainly concentrated in and around Chândnî Chawk. According to estimates more than 5,000 traders with more than 60,000 employees transact at least 10 % of the entire trade in textiles and cloth in India. Over 90 % of the cloth handled in Old Delhi comes from Ahmedabad and Bombay and is redistributed to Uttar Pradesh, Haryana, Punjab, Rajasthan, Himachal Pradesh. Only about one sixth of the goods find their final purchaser in the Delhi Region.

In spite of the efforts of urban planning authorities to relocate wholesale activities from the centre (Old Delhi) to the periphery of New Delhi, wholesale trade continues to grow in the Walled City and so does the demand for sales and storage space. The major economic incentive is the very low costs incurred in doing business and even manufacturing within Old Delhi, compared with other localities on the outskirts of Delhi.

Hence, the present day development of Old Delhi is characterized by commercial encroachment into traditional residential quarters and the transformation of historic structures. Almost all areas are threatened by the demand for commercial and storage space. This trend is paralleled by the outmigration of portions of the residential population.

The development of trade and manufacturing in Old Delhi is shown in Fig. 5. The number of enterprises involved in trade multiplied by seven during the two census decades 1961-1981; the number of manufacturing enterprises tripled in the same period. DDA estimates that the number of residents in the Walled City has been exceeded by the number of people in the daily working force since the mid-1980s. The extraordinary demand for industrial and storage space can generally only be met by converting and extending existing buildings. There are few examples of the construction of new multistorey commercial buildings (i.e. vertical extension) so far. Manufacturing and small workshops, in particular, are being relocated from the original bazaar areas to more peripheral residential areas. Here, new workshops are created by converting ground floor flats, storerooms and even courtyards.

This development is taking place in various residential areas regardless of their social status. Most affected, however, are quarters immediately adjacent to the major wholesale bazaars (Fig. 6).

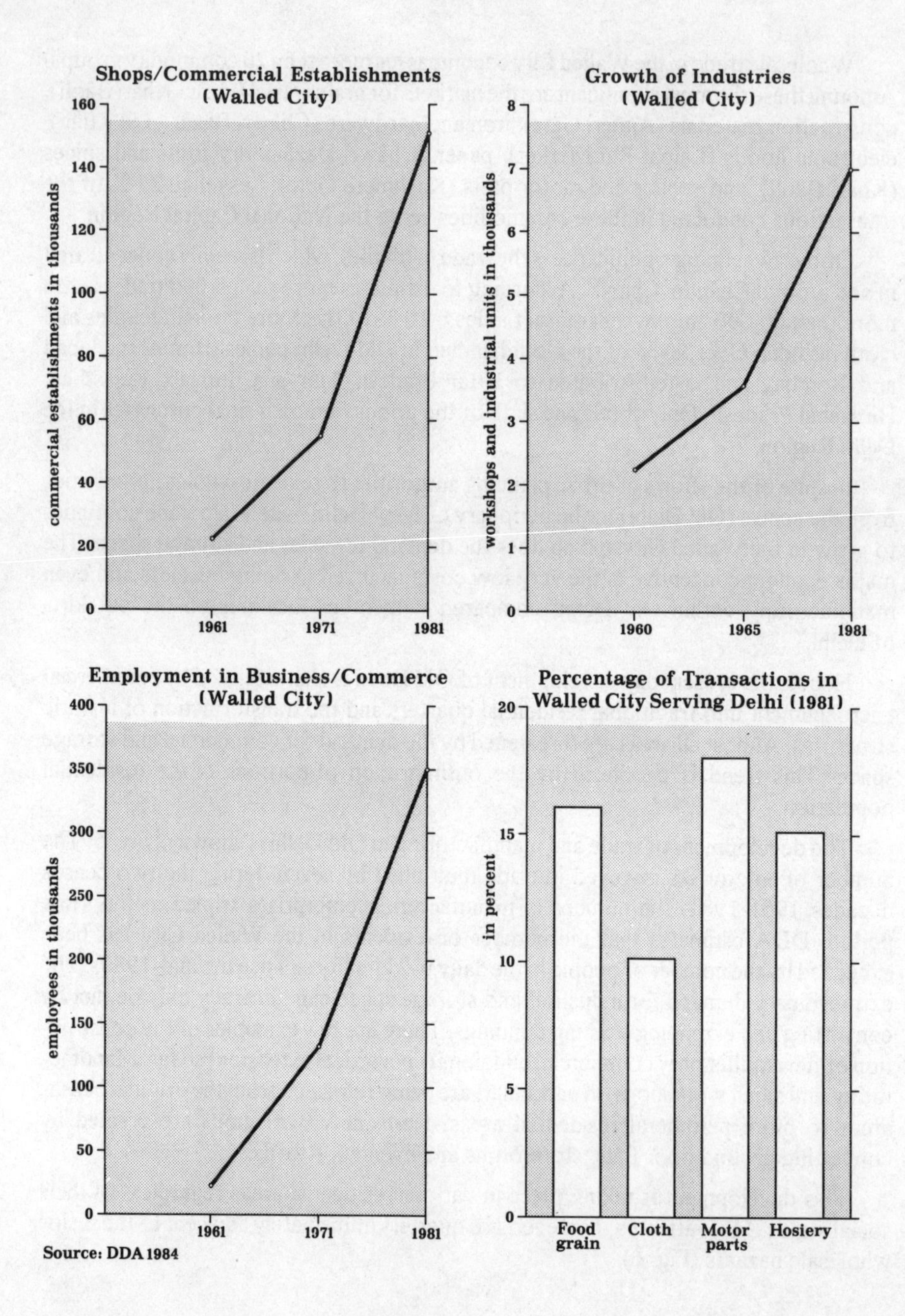

Fig. 5: Commerce and Industries in Old Delhi

The residential quarter Katra Neel (Katrâ Nîl) north of Chândnî Chawk, which until the end of the 1970s still had a "predominantly residential character" (Trivedi 1980, p.53) had become part of one of India's largest textile market by the late 1980s, with about one third of its total area serving this purpose. The mahallah, owing its name to the ancient trade in indigo, has traditionally been a Hindu residential area, especially for members of wealthy trader castes. Sanderson (1916, p.160f.) mentions several temples alongside the principal lane, next to two minor, rather insignificant mosques. A number of these temples, some of which date back to Moghul times, still exist today, mainly supported by income from religious endowments (such as rents from houses and shops).

Most of the Katra Neel residents belong to the middle class. In comparison to Walled City standards, the condition of buildings and infrastructure is good. Illiteracy is fairly low and there are three secondary schools in the mahallah.

According to its inhabitants, it was its high social status and the consolidated situation of property ownership that at first hampered the advance of commercial use. Since the early 1980s, however, the trend toward commercial infiltration and transformation cannot be overlooked in Katra Neel either.

Originally commercial use was almost entirely confined to the principal lane of the mahallah. Today shops and godowns[12] seem to be everywhere. New shops were constructed even in small by-lanes and few cul-de-sacs can be found that are restricted to housing only. The majority of the residents when interviewed claimed that the traditional mahallah structure was being destroyed by growing commercialisation. The influx of goods and non-residents into even the most remote side lanes changes the character of this originally semi-private space. For the residents this results in a loss of identification with their neighbourhood as well as in a general decline in housing quality. Anyone who has a chance of finding a new residence of an adequate social and financial standard outside this area is likely to migrate out of the Walled City at the first opportunity.

Demand for space has grown immensely due to the continous expansion of the textile trade. Hence enormous sums are being paid for the mere transfer of ordinary housing space to induce indecisive residents to move. This measure is re-enforced at times by massive pressure on tenants and deliberate damage to intact buildings while pulling down neighbouring houses to press for their demolition as well. In this manner the construction of new shops can be extended to the "cleared" area. According to DDA this planning zone showed a population loss of about 40% between 1961 and 1981. Recalculations from the 1989 electoral register show a population decrease of about 25% in the mahallah, compared with 1980.

Demographic changes in Old Delhi between 1961 and 1981 show distinctive spatial differences. Some planning zones have experienced a significant loss of population

12 godown = warehouse

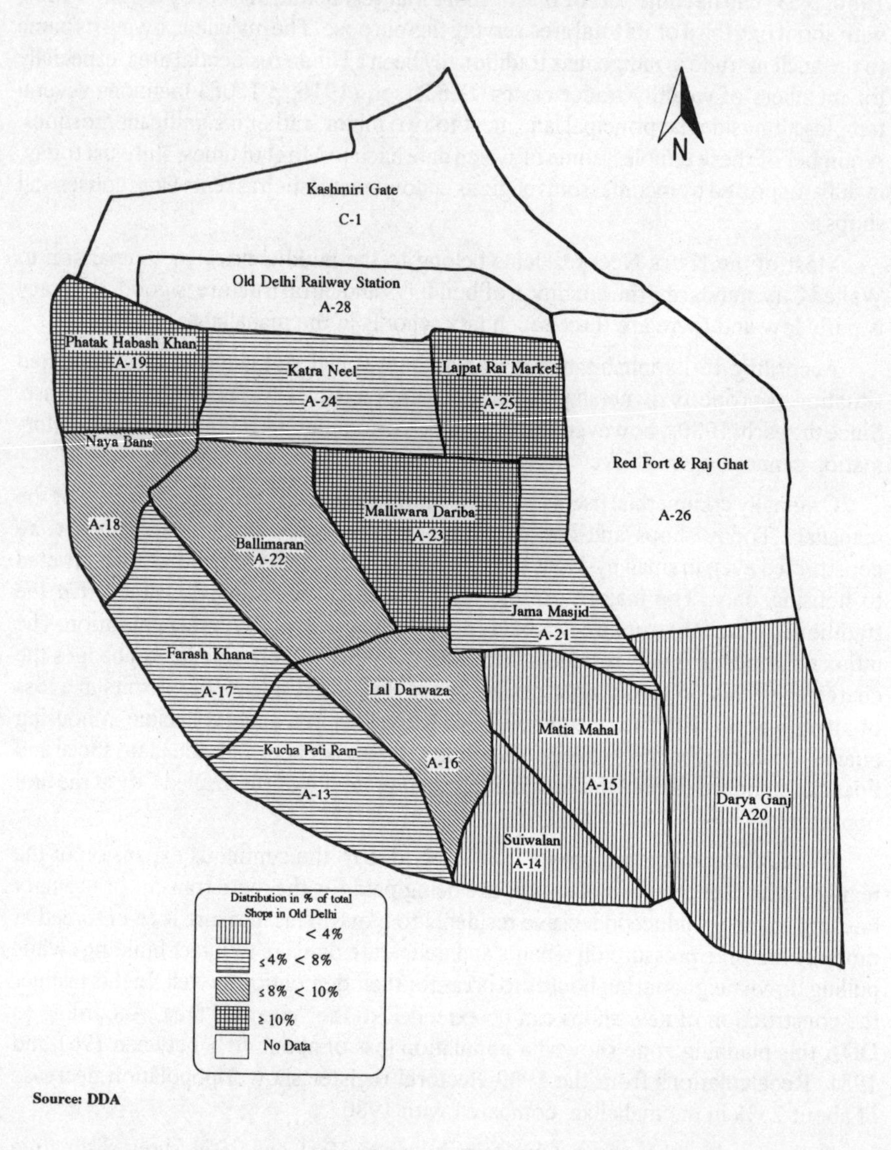

Fig. 6: Distibition of Wholesale Trades in Old Delhi

whereas others show a continous increase in population at the same time. Planning zones adjacent to the central bazaar areas show the highest losses. In contrast, considerable gains are registered especially in the predominantly Muslim wards in the southern part of the Walled City. The population of Darya Ganj with its good housing conditions and relatively high capacity for further population intake also increased during this period (Fig. 7).

Interviews conducted by the author in planning zones A-16 to A-24 in 1988 and 1989[13] as well as field studies conducted in 1989/1990 by geographers from Delhi University[14] in planning zones A-17, A-20, A-23 and A-24 provided insights into significant differences in migration behaviour of old town residents according to religious affiliation, socio-economic status and marital status. Some conclusions can be drawn from preliminary results:

Religious affiliation:

Hindus, Jains and Sikhs show comparable readiness to migrate. In contrast, there is a significant tendency toward residential persistence among Muslims. It may be assumed that this is partly due to the relatively weak socio-economic position of the majority of Muslims. Furthermore, all Muslim interviewees mentioned cultural identity and security[15] as decisive reasons for remaining in Old Delhi. Hence, Muslims prefer to move within the Walled City, a process resulting in a continuous population increase in Muslim quarters.

Thus one reason for the slight rise in the proportion of Muslims among the residents of Old Delhi lies in the specific migration behaviour of the community.

Socio-economic status:

Socio-economic status and distance between living and working place influence migration behaviour decisively. Members of the middle class make up a large majority of the people who are willing to move at the first available opportunity. Most of the affluent traders and manufacturers shifted their place of residence to New Delhi within

---

13 There were two sets of interviews conducted in 1988 and 1989, focussing on families still residing in Old Delhi.

14 Information by Prof. Aggarwal. These interviews included persons who had already migrated from the old town (contacts were established via remaining family members). The results have not yet been published.

15 Both sets of interviews were conducted under the impression of growing tension between the religious communities. Especially after the bloody unrests in Old Delhi in spring 1987 the factor "security" is overwhelmingly dominant for all interviewees. Muslims perceive the old town as a relatively safe place although it is the scene of most of the unrest.

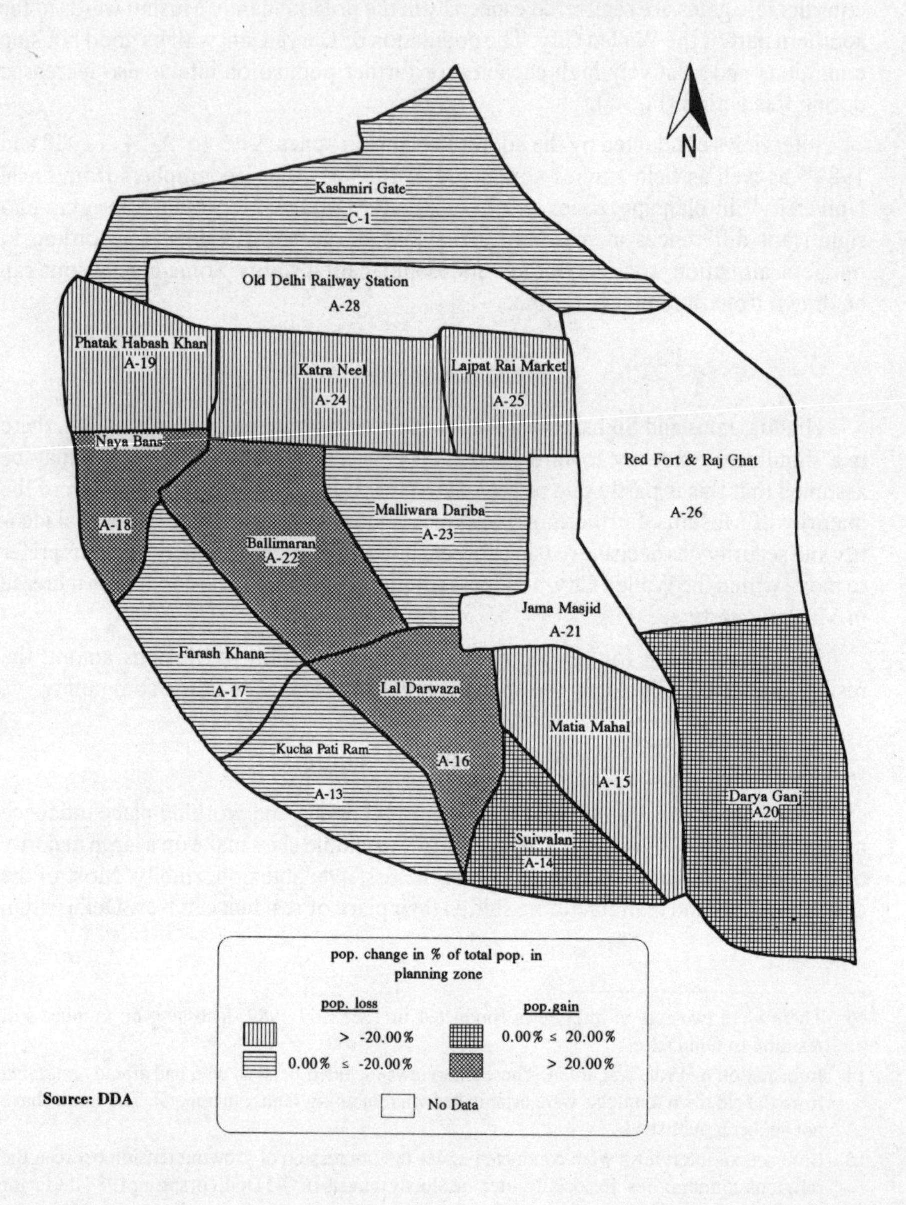

Fig. 7: Population Dynamics in Old Delhi 1951-81

the last three decades. Only the lower classes, unable to pay normal rents or daily fares to their places of work, are forced to remain in the Walled City. This results in a progressive social degradation of the old town quarters.

Marital status:

The break-up of extended family bonds is mirrored in migration behaviour. Marriage and the desire for a separate household, are the foremost reasons for migration. Hence the age group of 20 to 40 year olds comprises a disproportionate share of migrators.

A recent comprehensive survey of the socio-economic conditions of slum dwellers in the Walled City has revealed similar results (Kitchlu 1992). The study was undertaken to identify the problems being faced by the slum population in Old Delhi and their responses to redevelopment and/or resettlement options. Community living and protection and proximity to working opportunities are the foremost advantages identified by the respondents for living in Old Delhi. As with the former studies Kitchlu found a significant difference in the readiness to move betweens Muslims and Hindus/ Sikhs.

Building Fabric and Land Use

After massive and extensive interference with the building structure in the nineteenth and early twentieth centuries the contemporary development of the Walled City is characterised by progressive decay of the remaining traditional building substance:

> "The extreme congestion within the Walled City and incursion of a myriad variety of activities and trade have deteriorated the living conditions. Majority of the city's houses are in advanced state of decay and several areas have been overtaken by blight. During every monsoon quite a considerable number of houses collapse, but in spite of municipal warnings the occupants continue to live in the unsafe structures because there are not many places to shift. Every year scores of dwellings have to be pulled down by the municipal authorities as they pose danger to human life thus rendering many families homeless. According to a recent [...] survey 3,241 houses had to be demolished in 1964 as they had become totally derelict and unsafe for human habitation. Even otherwise a large number of dwellings have undergone radical transformation including additions and alterations which paradoxically made them more unliveable. Such houses rarely have adequate light or ventilation as they have been built up back to back with each other in a most chaotic manner without any plan or design."
> (TCPO n.d. p.15)

An overview of housing conditions in the Walled City can be derived from Tab. 1. The data are based on extensive investigations carried out by the Perspective Planning Wing of the DDA in the early 1980s. The account given in an internal Government report by the "Home Minister's Committee to Look into the Problems of the People Residing in the Walled City of Delhi" from 1987 differs, however. According to this report the majority (75 %) of the population living in the Walled City belongs to the economically weaker sections of society. Most households (73 %) consist of single room dwellings and about two third of the houses are rented.

Housing Conditions in the Walled City of Delhi

FACILITIES	in % of Housing Units
Kitchen	69.40
Bathroom	44.78
Flush toilet	37.99
Storage	22.13
Courtyard, indiv.	36.10
Water supply	77.43
ROOMS	in % of Households
Single room	50.04
Two rooms	33.43
Three rooms	9.32
Four rooms	7.20
OWNERSHIP STRUCTURES	in % of Households
Rental	54.0
Owner	46.0
MONTHLY RENT IN Rs.	in % of Tenant Households
0-25	57.37
26-50	22.15
51-100	11.18
100-200	3.49
Above-200	1.64
no data	4.17

Source: DDA 1984

Tab. 1: Housing Conditions in Old Delhi

The data on land use in Old Delhi are also contradictory. Tab. 2 shows the proportion of space dedicated to various forms of use in the early 1980s. As a result of the continous commercial encroachment into even remote residential areas it is almost impossible to distinguish between purely residential quarters and commercial and manufacturing areas.

Transport infrastructure takes up about one fifth of the total area. This is mainly due to the large railway facilities. The extension of the railways to the Walled City of Delhi brought new life to the city and played a crucial role in establishing the wholesale markets in Old Delhi during the late nineteenth and early twentieth centuries. Though roadtraffic has gained a considerable share of the transport market the railways are still an important factor for the location of wholesale business.

Land Use Area Analysis, Walled City of Delhi 1981

	Area in Hectares	Area in Percentage
Residential	181.01	31.86
Commercial	88.75	15.62
Manufacturing	9.59	1.69
Parks	90.42	15.91
Transport & utilities	131.51	23.14
Government & offices	18.74	3.30
Public	41.89	7.37
Others & vacant	6.31	1.11
Total	568.22	100.00

Source: DDA 1984

Tab. 2: Land Use in Old Delhi in 1981

More than 400 buildings, monuments or squares in Old Delhi are classified as historically important and worthy of conservation by the Archaeological Survey of India. To date, however, protective measures have only been taken for about 40 of them, most of them located inside the Red Fort. The rest (more than 350 objects) are endangered by misuse, deficient maintenance or real estate speculation. The idea of protecting not only single historic buildings but entire historic quarters in the Walled City has been accepted by most of the municipal planning authorities since the 1980s. Still there is a lack of financial means and legal instruments for carrying out even the most urgent measures. Hence there is no halt in the destruction of historic buildings and of social structures that have developed over a long period of time.

Rent Control and Land Market

Population changes following Partition resulted in substantial confusion regarding property ownership. Apart from this the Delhi Rent Control Act of 1958[16] is held responsible for the extensive decay of traditional building substance. Restrictions imposed on rent levels are said to keep owners' interests in preservation measures at a very low level. According to Maitra (1980, p.19) two factors account for the dilapidated state of the buildings in the Walled City. Firstly the very low rents paid by old tenants. The second factor is the present statutory provisions which mean that tenants cannot be evicted from their properties so making it impossible for owners to obtain the true value of their land. This leads to a very low standard of maintenance. In this way the rent control act creates a vicious circle as the owners are unwilling to invest money in their properties which would only then attract tenants willing to pay higher rents. A recent survey by the National Institute of Urban Affairs (NIUA 1989, p. 121 ff) concludes, however, that the effects of the Rent Control Act have by far been overestimated.

The combination of restrictions on rent levels (depending on duration of tenure), corresponding protection against unwarranted eviction, and a strict fiscal policy has created a system in the property and housing market characterized by fairly low rents, but immense sums to be paid as transfer fees. They have to be paid "illicitly" on an informal level and amount to as much as several lakh[17] rupees. If the owner knows about or even encourages the take-over of a rental object by a new tenant, he will usually pay part of this sum to the previous tenant. There are cases of previous tenants making a deal without the owner being aware of it. Formally the previous tenant will still be registered as tenant and will receive both the transfer fee and the difference between the actual and the newly fixed rent. This property market based on "illicit" transfer fees[18] is a result mainly of a strict fiscal policy. The Rent Control Act is of secondary significance only. Commercial take-over and transformation is promoted by these circumstances. There is no chance for administrative authorities or residents to defend their needs and interests on longer terms because they are confronted with enormous sums of money employed by the opposite side.

Jagmohan, the former Lieutenant Govenor of Delhi, describes the typical course of illegal transformation of housing space as follows:

---

16 The Rent Contract Act was amended in 1976 and then again in 1988. It regulates the fixation of a standard rent, tenant-landlord relations, evictions and lease of vacant premises to government organisations in the Union Territory of Delhi. The act was meant as a social measure to protect tenants against uncontrolled upward revision of rent and arbitrary evictions. However taken all the "exemptions and exclusions" of the Act into account only a small portion of Delhi's housing market is controlled by the Act.

17 lakh = 100,000

18 These transfer fees are generally known as "pugree" (pagaṛī).

"The manner in which unauthorized conversions take place is more revealing and significant than the construction itself. The process begins by closing doors of Katras from inside. The shops are constructed, through special contractors, in two or three days, working day and night. The lawyers are engaged before hand, and a scheme of thwarting the law is prepared. When the municipal authorities serve notices under the corporation Act, 'stay-orders' are obtained from the civil court, claiming that the construction is old. During the period of 'stay-orders' the walls of the Katras are broken, and the shops are opened on the main street and lanes. In accordance with a preconceived plan, these shops are sold clandestinely after obtaining under the table, huge amounts in the shape of pugree. All the transactions take place in black-market. After occupation, the individual burger of the unauthorized shops also obtains 'stay-order' from the civil court. All means are employed to prolong the court proceedings. Meanwhile, pressures are generated. Full advantage is taken of the tendency of the officers to follow the line of least resistance. Ultimately, unauthorized markets come to stay. In some cases, as large as Rs. 25 lakhs are obtained as pugree. Hundreds of such cases have taken place and many more are taking place. But in no case, the offender has been brought to book."
(Jagmohan 1975, p.35f)

Conclusion

The scale and extent of business activities in the Walled City have multiplied in the last decades causing a general transformation of the historic city. Former residential areas have been completely invaded by commercial establishments, resulting in population decrease. Out-migration is influenced by social, religious and marital factors and is changing the overall composition of the population residing in Old Delhi.

Recent studies on the traditional city core of Bombay (Sita 1988) and of the Walled City of Hyderabad (Naidu 1990) have shown similar demographic and functional changes and indicate that the processes described above are of a more general nature for Indian cities.

Inspite of the attempts over several decades in town renovation and conservation planning living conditions in Old Delhi have constantly worsened. The relocation and resettlement policy favoured for so long by planners and politicians has to be seen as a total failure (cf.: Ali 1990; Misra 1981). In particular the slum clearance measures carried out during the Emergency, have left the residents of the Walled City with a massive distrust of any form of intervention by the planners.

Even today the planning authorities have no comprehensive concepts for the conservation of the building fabric and social structures in the Walled City. However, even if such concepts existed their implementation would be plagued by severe difficulties. Particularly the influential trading community of the Walled City has so far successfully resisted all attempts to relocate the wholesale trade from the old city to the periphery of New Delhi.

The few successful relocation projects have only led to the provision of urgently needed space for the merchants remaining in the Walled City. For the wholesale merchants a relocation from Old Delhi would threaten the close network of interrelationships within the bazaar and also mean the loss of enormous investments in the form of pugree and in real estate. Due to its significant influence on politicians and officials, the Old Delhi trading community has proved successful in defending its own interests. It seems therefore doubtful that the process of deterioration of the traditional residential areas of the Walled City caused by indiscriminate commercialization, overcrowding and traffic congestion is likely to be stopped in the near future.

Bibliography

Akram Rizvi, S.M.: Kinship and Industry among the Muslim Karkhanedars in Delhi. In: Family, Kinship and Marriage among Muslims in India. Edited by Imtiaz Ahmad. New Delhi: Manohar. P. 27-48

Ali, Sabir: Slums within Slums: A Study of Resettlement Colonies in Delhi. New Delhi: Vikas Publishing House 1990

Ansari, Iqbal A.: The Muslim Situation in India. New Delhi: Sterling Publishers 1989

Aziz, Abdul: Changing Face of Delhi. A Geographical Perspective. Aligarh: Department of Geography Aligarh Muslim University 1983

Bharat Sevak Samaj, Delhi Pradesh: Slums of Old Delhi. Report of the Socio-Economic Survey of the Slum Dwellers of Old Delhi City. New Delhi: Atma Ram & Sons 1958

Commitee on Reorganisation of Delhi Set-up. Report Part I & II. New Delhi: 1989

Datta, V.N.: Punjabi Refugees and the Urban Development of Greater Delhi. In: Delhi Through the Ages. Edited by R.E. Frykenberg. New Delhi: Oxford Univ. Press 1986

DDA: Walled City Shahjahanabad. Planning Issues and Policy Frame. A Report of Perspective Planning Wing. New Delhi: 1984

DDA: Master Plan for Delhi. Perspective 2001. Modified Draft. New Delhi: 1987

DDA: The Future of New Delhi. Proceedings of the Seminar on the Future of New Delhi. New Delhi: 1984

DIT: Administration Report of the Delhi Improvement Trust for the Years 1939-1941. New Delhi: Latifi Press 1942

Evenson, Norma: The Indian Metropolis. A View Toward the West.New Delhi: Oxford University Press 1989

Ghosh, Bijit ed.: Shahjahanabad: Improvement of Living Conditions in Traditional Housing Areas. Proceedings of the Seminar: Improvement of Living Conditions in traditional Housing Areas (23.-29. February 1980); Goethe Institute, New Delhi. New Delhi: 1980

Goodfriend, Douglas E.: Changing Concepts of Caste and Status among Old Delhi Muslims. In: Modernization and Social Change among Muslims in India. Edited by Imtiaz Ahmad. New Delhi: Manohar 1983, p. 119-152

Gupta, Narayani: Delhi between Two Empires 1803-1931: Society, Government and Urban Growth. Delhi: Oxford University Press 1981

Gupta, Shiv Charau: Delhi: The City of Future. New Delhi: Vikas Publishing House 1987

Jagmohan: Rebuilding Shahjahanabad: The Walled City of Delhi. Delhi: Vikas Publishing House 1975

Jain, A.K.: The Making of a Metropolis. Planning and Growth of Delhi. New Delhi: National Book Organisation 1990

Kitchlu, T.N.: Survey of Evacuee Properties in the Walled City of Delhi. New Delhi: Delhi Univ. Dept. of Social Work 1992

Mehra, Ajay K.: The Politics of Urban Redevelopment: A Study of Old Delhi. New Delhi: Sage Publications 1991

Misra, Girish and Gupta, Rakesh: Resettlement Policies in Delhi. New Delhi: Indian Institute of Public Administration, Centre for Urban Studies 1980

Mohan, Inder: The Jama Masjid Controversy. In: DESIGN 17(1973), no. 3, p. 23-29

Naidu, Ratna: Old Cities, New Predicaments. A Study of Hyderabad. New Delhi: Sage 1990

National Capital Region Planning Board: Regional Plan 2001. National Capital Region. New Delhi: Ajanta 1988

National Institute of Urban Affairs: State of India's Urbanisation. New Delhi: 1988

National Institute of Urban Affairs: National Capital Region. A Perspective of Patterns and Processes of Urbanisation. New Delhi: Research Study Series 29. 1988

National Institute of Urban Affairs: Rental Housing in a Metropolitan City. A Case Study of Delhi. New Delhi: Research Study Series 37. 1989

Oldenburg, Philip: Big City Government in India: Councilor, Administrator and Citizen in Delhi. First published in the USA 1976. New Delhi: Manohar 1978

Pillai, Lakshmi: Decision Making in a Public Organisation. New Delhi: Manohar 1991

Rao, V.K.R.V. and Desai, P.B.: Greater Delhi: A Study in Urbanisation 1940-1957. Bombay: Asia Publishing House 1965

Sanderson, G.: List of Muhammadan and Hindu Monuments. Volume I: Shahjahanabad. Calcutta: Superintendent Government Printing 1916

Sita, K.: The Declining City-Core of an Indian Metropolis. New Delhi: Concept Publishing 1988

TCPO: Redevelopment of Shahjahanabad. Working Paper. New Delhi: 1974

TCPO: Redevelopment of Shahjahanabad. The Walled City of Delhi. Resume of the Seminar. New Delhi: Government of India Ministry of Works and Housing, n.d.

Trivedi, H.R.: Housing and Community in Old Delhi. New Delhi: Atma Ram 1980

Plans of Indian Towns

An outline of the extant maps and plans of Indian towns, up to the middle of the 19th century, with special reference to maps of Delhi.

Susan Gole (London)

Similarities between the layout of several towns of the Harappan civilisation are so great that one might suppose a common plan had been drawn up, and sent out to new colonists. Unfortunately, if such a plan ever existed, there is no trace of it now. In fact, there is very little trace of any such plans made before the eighteenth century, only stray references in the literature to what might have been plans of battles, incised ground plans for the construction of a temple on the nearby rocky hillside, and symbolic plans of pilgrimage towns. A few of these pilgrimage maps may date from the seventeenth century, but since their style has not changed much even today, it is not easy to know when they first became popular.

Early town plans

In Europe, the town plan developed over many centuries. In his Introduction to the facsimile edition of Braun and Hogenberg's Civitates Orbis Terrarum, R.A. Skelton wrote that 'during the Middle Ages towns were usually represented in profile, as seen from the ground, with emphasis on conspicuous or important buildings. The study of the science of perspective in the 15th and 16th centuries developed the birds-eye (or oblique) view taken from a higher point of vision, as exemplified in the landscape backgrounds of many oil-paintings...In the middle decades of the 16th century, however, geographers, surveyors and engineers introduced geometrical methods of mensuration enabling smaller areas, as in a map, to be delineated from a vertical viewpoint. So the town plan was reborn, and it quickly became recognised as the only form of topographical representation in which spatial relationships were correctly expressed and the structural growth of the city, in terms of history and geography, could be discerned'.[1] This is only partly applicable to India, as will be seen

1 Published 1964 (World Pub. Co., Cleveland and New York). This was the first large collection of town plans (1572), published over many years as material was gathered, and reprinted up to 1657. Only four towns in India are represented, all Portuguese settlements, Calecut, Cananore, Diu and Goa, copied from rough sketches supplied by a Hanse merchant Constantin von Lyskirchen.

later. Some styles of mapmaking certainly developed independently in several parts of the world simultaneously or at various periods. Till more is known about Indian cartographic development it is not possible to determine exactly how the growth of the town plans in this part of Asia fits with European models.

Town plans are known from China from a much earlier date. Carved on stone in 1229 A.D. is a detailed plan of Suchow in which pictorial details—important buildings, the city walls, the hills outside—are superimposed on an outline ground plan.[2] Early plans were also made for pilgrims going to Mecca, though few have survived even from the 16th century. One example of an early Mecca plan is known from India, made in 1676[3]. Another influence might have come through Persia from Turkey, where detailed town plans were made for the description of Sultan Sulaiman's travels in the 16th century.[4] So there were examples perhaps known in India from early times, from which the idea of detailed town plans might have come, and more may come to light. Describing a Portuguese plan of Goa by Pedro Barreto de Resende made in 1646, James Elliot thought that Resende was 'influenced by native Indian cartography, as the use of thick vivid colour and the crowding together of groups of stylised minor buildings may testify'. Apparently the publisher accused Resende of executing the plans without proper scientific measurement, perhaps finding this style somewhat old-fashioned and oriental.[5] Unfortunately no Indian model made prior to this date has survived, so their existence can only be postulated.

Early European Plans of Indian towns

The earliest European plans of Indian towns were made to illustrate the Portuguese settlements. In fact, looking at collections of town plans made in the late 16th and early 17th centuries one might be forgiven for thinking that there were no substantial urban developments in India before the arrival of the Europeans. Though there are descriptions of the extensive town of Vijayanagar by European travellers, no plans were available for the cartographers compiling their atlases and travel books back home. The same is true for the capital of the Mughal empire, Agra. The first detailed

---

2 P.D.A. Harvey, The History of Topographical Maps: Symbols, Pictures and Surveys, London: Thames and Hudson, 1980, p.107. This book supplies the most detailed study of the development of town plans to the end of the 17th century.

3 Sadashiv Gorakshkar, 'An illustrated Anis al-Haj in the Prince of Wales Museum, Bombay' in R. Skelton et al (eds), Facets of Indian Art, London, Victoria and Albert Museum, 1986, p. 158-67 (Reproduced in Susan Gole, Indian Maps and Plans (hereafter IMAP), New Delhi, Manohar, 1989, p. 163).

4 Huseyin G. Yurdaydin, Nasuhu's-Silahi (Matrakci): Beyan-i Menazil-i sefer-i irakeyn-i Sultan Suleyman Han, Ankara: Turk Tarih Kurumu Basimevi, 1976.

5 James Eliot, The City in Maps: Urban Mapping to 1900, British Library, 1987, pl. 4. The map of Goa was published in Antonio Boccaro, Livro do Estado da India Oriental, 1646.

map of Goa, drawn from observation, was published by Jan Huygen van Linschoten in 1596, who spent seven years there.[6] In 1672 Phillipus Baldaeus published several plans and views, but again only of towns where Europeans had established factories.[7] As the British too gained control of the towns where they had factories, maps began to appear in publications. The first map of Bombay was drawn by John Ovington in 1668, but not published until 1696. John Fryer also brought back a rough sketch in 1698, and John Thornton included a map of Bombay and the island of Salsett in his atlas of sea charts in 1703. The first detailed chart of Bombay harbour did not appear until 1763, surveyed by William Nichelson, and it showed very few inland features.[8]

Other British towns were also surveyed as British power spread. In the Fort St George Museum, Madras, is 'A Prospect of Fort St George and Plan of the City of Madras Actually Surveyed by Order of the Late Governor Tho. Pitt Esq' made by John Harris and John Friend, (Pitt was Governor 1698-1709) and there were many plans of Madras in European publications, especially during the mid-18th century when the French and British struggles were at their height. From French sources came maps of Pondicherry, the first to be published probably that by Nicholas de Fer in 1705, and a plan of the Dutch mission at Tranquebar was drawn in 1709. Plans of Calcutta were made for Orme's account of the military campaigns published in 1763, and the first detailed survey was made by A. Upjohn in 1792-93.

British Surveys

By the end of the 18th century plans of most of the towns that appeared in reports of fighting in India had been made for the news-hungry readers in Europe. Gradually as the Europeans dispersed over the country and settled in new areas, more plans were made. The first Surveyor General of Bengal, James Rennell, was appointed in 1767, but most of his work was on a smaller scale than that required for town plans. The start of the triangulation surveys in 1802 soon led to the inauguration of the Revenue Surveys, and with them detailed plans of areas and towns on a large scale. It was, however, a haphazard approach, with maps appearing of certain towns and districts at irregular intervals depending upon the penetration of the survey teams, the

6 In Itinerario, voyage ofte Schipvaert, Amsterdam: Cornelis Claesz, 1596 and editions up to 1610; see Susan Gole, India within the Ganges (hereafter IWTG), New Delhi: Jayaprints, 1983, p. 113, and Van der Aa's copy reproduced in S. Gole, A Series of Early Printed Maps of India in Facsimile (hereafter EPMIF), New Delhi: Jayaprints, 1984, No. 26e.

7 In Naauwkeurige Beschryvinge van Malabar en Choromandel. Amsterdam: J.J. van Waasberge & J. van Someren, 1672. See IWTG, p. 127-8.

8 John Henry Grose's plan of Bombay is reproduced in IWTG, p. 184, and that of Francois Bellin in EPMIF, No. 33e. De Fer's plan of Pondicherry is reproduced in EPMIF, No. 23b, and Moll's plan of Fort St George in IWTG, p. 83. Kitchin's Territory of Calcutta from Orme's A History of the Military Transactions of the British Nation in Indostan, Vol. I, is reproduced in IWTG, p. 178, and an earlier one by Francois Valentyn of 1725, ibid, p. 70.

willingness of local rulers to admit them, the finances of the Company, and the capabilities of the surveyors. Naturally there was more demand for areas under the direct control of the British, and where they collected the revenues themselves, than for those areas where there was lesser involvement. Very often the main system of roads in the towns can be made out from smaller scale maps of the whole district, but no other detail was supplied.

Most of the major towns of British India were not fully surveyed until the 3rd decade of the 20th century, and even today there are many large towns without accurate maps. But then, on the whole, people in India have rarely needed maps either to traverse the country on pilgrimage, as they have done for centuries, or to go from one part of town to another. There was always someone who knew the way, had been there before, or was able to start the traveller in the right direction so that he could obtain further information as he progressed. Since so few maps made in India have survived in collections in India or abroad, it was assumed that there was no knowledge of cartography in India at all till the coming of the Europeans. Lately, however, some two hundred maps and plans of various types have been noted, and possibly more will now be located.[9]

Indian town plans

Religious maps - Jain

The town maps that are Indian in style can be divided into two categories, the religious or pilgrimage maps, and the purely topographical. Most of the pilgrimage maps appear to have been made as souvenirs to be carried home, rather than as guides to the sites to be visited. Among the most popular of the Jain holy places are Parasnath Hill, Mount Abu, Palitana, and Girnar, and paintings of all these towns have survived. Some are definitely views, others are more map-like. An 18th century pata (cloth painting) of Shatrunjaya in Gujarat [1] shows the different routes open to the pilgrim as he progressed from temple to temple up the hill to the summit, leaving the town of Palitana below. There is no sense of scale, since the purpose was to depict the holy sites, and other geographical detail was irrelevant. Another painting, probably 20th century though perhaps the style is much older (and this may have been a copy) is of the Shri Nathji temple at Nathdwara [2]. Here there is far more accuracy in the depiction of each building, even to the doors and windows on each house, yet the style remains distinctly Indian.

Another type of Jain map which depicted towns, even buildings, in great detail was the *Vijnapatipatra* [3]. These were invitations to Jain pontiffs to spend the monsoon in a particular town, thereby blessing it with their presence. In the form of long scrolls,

9 Most of those that have surfaced so far have been illustrated in Susan Gole, Indian Maps and Plans (hereafter IMAP). New Delhi: Manohar, 1989.

they often depicted both the towns the monk would pass through, and the host town, as a form of enticement to the delights on the way. Their style owes more to folk-art than cartography, but they give an idea of the respective location of important buildings and their size.

Religious maps - Hindu

Among Hindu maps are the towns of Vraj [4], Jagannath [5], Ujjain [6], Kashi [7-9], Sankhodar Bet [10], Dwarka [11] and an unidentified temple and town on an island [12]. They are in many styles, no two alike, each one placing emphasis on different details, and thus drawn with a different aim by the cartographer. Kashi, or Varanasi, is represented by three types of maps. One [7] is a modern pilgrim's map still on sale in the bazars today, crudely drawn and brightly coloured. It shows a circular town within a wall, crowded with buildings and temples, with blank space filled by stylised trees. Another [8] is a maunscript painting on paper, with Sanskrit text, to show the five-cos circumambulation of the city. The Ganges with its tributaries the Asi and Varuna form a prominent feature. The only other topographical details are the temples to be visited, each one named, and trees on the river banks. The third map [9] is a large cloth painting, filled with detailed depictions of buildings, and crowds of pilgrims making their way along the seven paths around the holy city. The map is in poor condition, and so far no attempt at identifying the various topographical features has been made. The town is in the form of a semi-circle, with the Ganges flowing along the base.

Wall paintings

Town plans or views are frequently found as wall paintings, especially in Jain temples. No old ones have been found, as the walls are frequently repainted, and the images again replaced, usually as they were originally depicted. They form, in a way, a guide to the pilgrimage that each orthodox Jain should hope to complete. No order in their placing around the wall has been determined.

Another location for wall paintings of towns is in the Bhojan Shala at the palace of Amber, near Jaipur [13]. This was the private dining-room of the Raja, and around the wall are paintings of five holy towns. Contemplation of these sites was considered an aid to digestion. No date has been assigned to these paintings, but it can be presumed that they were made before the move to the new town of Jaipur in 1727. Similar paintings have been found on the walls of some houses in Shekawat, a district of Rajasthan, by Ilay Cooper,[10] though the surviving examples are certainly of more recent date. Others have been recorded in Bombay.

---

10 Ilay Cooper, The Guide to the Painted Towns of Shekhawati, Churu (Rajasthan): Girish Chandra Sharma, [1987].

The Jaipur collection

Jaipur, in Rajasthan, stands out as a centre for map-making, or perhaps it is the only place where a large number have been preserved. More than 30 maps are in the City Palace Museum, ranging from large cloth paintings to small plans on paper. Some were clearly drawn on the spot, as plans for construction, others, such as the large plan of Surat [14], seem to have been brought from elsewhere. A large (292 x 272 cm) map of Agra [15]—either unfinished or drawn to show certain elements only—shows main roads within the town wall, gardens along the banks of the river Jumna, with a few areas sketched in more detail. Notes on the map indicate that it was made at a time when Sawai Jai Singh II was governor of Agra, in the middle of the 18th century, to show what repairs were required in the defences, marked yellow to differentiate them from the rest of the wall. Some buildings are also shown in elevation.

An even larger map (645 x 661 cm) shows the town of Amber in great detail [16]. It has been tentatively dated to 1711, before the new capital of Jaipur was built in 1727. Every house on every street is drawn and named according to the occupation of the inhabitants. It includes the outlying areas and the hills around, with every spur and valley clearly shown. The buildings are sometimes drawn conventionally, and are in both plan and elevation. It is a remarkable record, unique in its clear depiction of a town soon to be abandoned. Nothing is known of its maker, nor the reason for which it was drawn, but it is obviously the work of extensive surveys which must have taken some time to complete.

The map of Surat already referred to [14] is in a very different style. Also on cloth but slightly smaller in size, it depicts a regular geometric pattern of streets, with the name of the occupants of each block written in Persian. A rough translation in a Rajasthani dialect has been added. Around the edge of the map outlying villages have been shown conventionally, but with no recognition of their respective distance from Surat. No small streets or lanes are shown, so it might be conjectured that the person for whom the map was made, or for whom it was purchased, wanted to locate only the foreign factories and wharfs—identified by their flags, the residences of important Mughal nobles, and certain public buildings. Other information was irrelevant. Perhaps such maps were available, and since this example has survived in Jaipur, and the translation from Persian has been added, it might have been taken back by Jaipuri agents, or even ordered specially by the Jaipur court. No other copies are known, though Surat [17] is also depicted on a small manuscript account of a pilgrimage to Mecca. Here the important features shown prominently are the mosque in the town, and the relationship between town and fort and the harbour where pilgrims embarked for the voyage across the Indian Ocean. The style is reminiscent of earlier plans of Mecca itself, which have survived in some numbers.

A large map of Srinagar in Kashmir [18] has also been preserved in the palace at Jaipur. Again on cloth, this one has no text at all, though east is indicated by the rising sun appearing from behind the snowy mountains. The Dal lake occupies a large area, with the formally laid-out Mughal gardens along its shores. Across the lake is the fort of Hari Parbat, with the town spread out beyond. The bends of the Jhelum river and its many tributaries have been shown in considerable detail. Roads are indicated by a narrow red line, and within the town the blocks of buildings appear to have been placed from observation, not convention. Srinagar is the only town in India of which several early maps survive. Whether it attracted more map-makers, because it was easy to climb either Shankaracharya Hill or Hari Parbat, and see the town spread out below, or whether it is this group which has chanced to survive, remains at present an enigma.

From the neighbouring state of Jammu there is a single map of the town in the bend of the Tawi river [19], dominated by the palace. This map is drawn in such fine detail, perhaps during the time of Ranjit Singh or later, that it should not be hard to date it, or compare the lay-out of streets with those of today.

Another map which appears to have been constructed from personal observation if not actual survey, shows the town of Nasik [20]. No detailed study is known, but it seems to have been made in the third quarter of the 18th century. There are references to Naoroshankar who was governor of the city on behalf of the Holkars in the 1760s. No European plan, of which this might have been a copy, is known, yet the style stands out among indigenous maps in its meticulous lines and detail of individual buildings. It also, however, retains something of the geometrical figures of the Surat map. Even today, no large-scale plan of Nasik is available with the Survey of India, so it would be fascinating to discover the origin of this very detailed map.

From further south, plans have survived of Bijapur [21] (one dated perhaps to the end of the 17th century, certainly not before 1686 AD), Udgir [22], Dharwar [23], Janjira [24], Vijaydurg [25], and Hyderabad [26]. Each one is different in style, and there is no common thread of development running through them. Clearly the objectives for which they were made varied, so the emphasis is thrown on different aspects of the town. Of Dharwar [23], only the fort is shown in detail, but rude outlines of the surrounding town have been appended on pasted slips of paper on three sides of the map. That of Janjira [24] depicts one of the battles fought by the Marathas in their attempt to dislodge the Sidis between 1758 and 1761, and gun emplacements are prominently depicted between elevations of important buildings. In contrast the large plan of Hyderabad [26] (possibly dated to 1772 AD) shows a wealth of social detail. From it one can see the styles of domestic architecture, modes of conveyance, the clothing of different classes of the many people depicted on the streets. There is no text on this map, except a single toponym locating Daralshifa, a hospital. For whom was it made, and who was the maker? Were there others? The mysteries remain unsolved, but one can say that it is unlikely this map ever left Hyderabad where it can now be seen.

Plans of Delhi

Indian plans

Having discussed the growth of town plans in general, and the few that have survived which belong to a peculiarly Indian style, I turn to maps of Delhi itself. The city now known by this name was built by the emperor Shah Jahan north of the ruins of several earlier cities near the same site. No construction plans are known to have survived, though there is a literary reference to the emperor commenting on the covered arcades of the markets at Isfahan, and saying he would like to have them copied in his new capital. Royal interest lay in the construction of the fort itself, while the surrounding plots were taken over by nobles and merchants, with those lower down the social scale fitting their dwelling places into the space that was left. The city wall was added towards the end of the 17th century.

As Rory Fonseca has pointed out, 'the city assumed its final shape around six important architectural and planning elements'.[11] These were the (i) Fatehpuri Masjid erected in 1650 one mile due west of the palace's Lahore Gate; (ii) the gardens to the north of the pathway (iii) leading from the palace to Fatehpuri Masjid, known as Chandni Chowk; (iv) the Jamma Masjid built on a rocky outcrop southwest of the palace; (v) the Faiz Bazar, the second most important street, running south from the Delhi Gate of the palace to the Delhi gate which led to the old town of Delhi on the south side. The sixth nodal element was the main water reservoir at Hauz Qazi, situated at the junction of four important bazaars.

The earliest surviving map of any part of the Mughal town of Shahjahanabad is probably the small plan of the Red Fort in the Jaipur collection [27]. Sketched on a hand-drawn grid in red and black it shows and names the main buildings within the fort wall and the bridge across to Salim Garh. On the right is a pencil outline of the house of Ghazdi Khan (of whom no more is known). The style is similar to a plan from the same collection of the fort at Agra [28]. They were perhaps made in the middle of the 18th century.

The fort features on another plan, known from two surviving copies, which appears to belong with detailed plans of the two main streets, Chandni Chowk and Faiz Bazar [29]. Here the fort is shown from the outside only, with the interior left blank. The two street plans are drawn as they might be viewed by someone walking along the centre of each street (Plate 1). Buildings on either side of the road are shown in elevation, with some taller buildings obviously visible from the road above the arcades of shops on either side. Even the trees beside the central canal seem to have been drawn accurately, though this of course cannot be proved. The occupant of each house is named, as well as the important public places such as the baths of Sa'ad-ullah Khan, mosques, the Kotwali, and the gate to the Begum's garden. The text is Persian, with

11 Rory Fonseca, 'The walled city of Old Delhi' in Ekistics, 31, no. 182, January 1971, pp. 72-80.

the addition of French translation on one of the maps. They too appear to date to the late 1750s before the destruction of the city by the Afghans and the Rohillas.

The town of Delhi also features on two large maps from about this period. One is a very detailed map of the Ali Mardan Khan canal from its start near Benawas to the outflow into the Yamuna south of Delhi [30]. Various bridges over the canal are named, as well as the residences of the nobles along its banks outside the city wall. The way the Sa'ad-ullah Chowk and Faiz Bazar are depicted is very similar to that on the two street plans named above, and they are possibly from the same date. The map is in the form of a long scroll, about 12 1/2 metres in length, so there has been no attempt to show orientation or scale. The second map [31] is also a scroll, known in two copies (25 x 2000 cm and 20 x 1200 cm). It depicts the road from Shahjahanabad to Kandahar and was probably drawn later than the canal map, since many of the villages are noted as being in ruins. On the other hand there is a reference to Raja Amar Singh of Patiala who died in 1765. The main roads and bazaars of the city are drawn and named, but again, being a scroll map, there is no attempt at orientation, yet it would have served well as a guide for a visitor coming to the town for the first time. It is certainly depicted in more detail than any of the other towns on the road from Kandahar.

One more map of Delhi [32] appears to be from an Indian hand, though it is known only from a copy (Plate 2). The English copyist has styled it 'Plan of Dehly Reduced from a large Hindostanny Map of that City' and someone has added '1800 (?)'. It shows a square plan within the wall which encloses also the fort. Apart from the gates and measurements of the distances between the bastions on the wall, the only names are of Safdar Jang's house, Ishmael Khan's house, and Jama Masjid. The square shape has meant some distortion to the layout of the streets, particularly in the south west quarter. Faiz Bazar is shown as only a third as long as Chandni Chowk, and Jama Masjid is much too close to Turkman Gate, leaving little space for the maze of streets south of Chawri Bazar. This important commercial centre is shown as wide as its name implies, from the mosque to Hauz Qazi, mirroring Chandni Chowk from the fort to Fatehpuri Masjid.

European maps of Delhi

The earliest European map of Delhi that has been traced appears to be the one possibly acquired by Colin Mackenzie during his single visit to Delhi in 1814-15 since it was formerly catalogued as no. 35 in his collection of papers at the India Office Library [33]. It is titled 'Trigonometrical Survey of the Environs of Delhy or Shah Jehanabad. 1808'. Its size is 24 x 18 inches, and its scale 1 inch to the mile. No surveyor's name is mentioned, nor any record of it having been printed. It covers a much wider area than the city itself, but the layout of the streets in relation to the fort and the city wall is fairly accurate.

Another map [34] possibly made soon after that acquired by Colin Mackenzie is probably the one that was hurriedly zincographed in England on 3rd August 1857, when the British forces were planning to recapture the town from the rebellious sepoys. The printed title states: 'Plan of Delhi. Copied and Zincographed at the Ordnance Survey Office, Southampton, from a M.S. Plan in the possession of Col. Edward Harvey. Unattached. August 3rd 1857.' Phillimore[12] has recorded an entry in a fieldbook about 'a most interesting plan of the city, 200 yards to an inch, signed by Peter Lawtie 13th December 1812, which shows the old city gates, including the *Kashmere Gate* and also the *Koorseah Bagh*, and the road outside the city wall'. The large scale of 200 yards to the inch suggests that he is here referring to the same map that is preserved in the India Office Library but dated 1857.

Delhi is depicted in small scale on sheets of the Revenue Survey surveyed by Captains T. Oliver, I.H. Simmonds and W. Brown during the years 1824-32 and 1840 [35]. They were compiled and reduced in the Surveyor General's Office, Calcutta in August 1848 on a scale of 4 miles to the inch. The outline of the city is more similar to the earlier square 'Hindostanny' map and the interior roads in the south-west quarter are shown with an orientation of west-east instead of the correct northwest-southeast. The large survey by F.J. Burgess [36] titled 'A Topographical Survey of a Portion of the Dihlee District shewing the principal Basins of Drainage, and the positions of the Old Bunds, also the Situations of the Ancient Cities and the Objects of archaeological Interest' gives a fairly detailed though small depiction of the town which he calls 'Shahjahanabad or Modern Dihlee'. The scale is also four miles to the inch. This was an influential map, copied many times, and was probably the basis for most of the maps published during the conflict of 1857-58 and later. Two versions are preserved in the India Office Libary [37], both quite small since they show only the town and its immediate neighbourhood, and are titled 'The Fort and Cantonment of Delhi', with acknowledgement to Burgess's survey.

12 R.H. Phillimore: Historical Records of the Indian Surveys, Dehra Dun, 1946-52, II, p. 61.

Tab. 1

left: Chandni Chowk on a manuscript of c. 1756 [29]. At one end is the mosque built by Fathpur Begum, mistakenly shown with three domes. Persian text names the owners of many of the *havelis*, and at some time a French translation has been added. The buildings on either side are delicately drawn, though small in relation to the street itself

right: Faiz Bazar, on a manuscript similar to that of Chandni Chowk [29]. The Akbarabadi masjid on the west side of Chowk Sa'adullah Khan was later pulled down, but the smaller one next to it was allowed to remain. At the other end of the street is the Delhi Gate, leading to the older cities of Delhi south of Shajahanabad. The fact that the canal ran through the centre of this street too is rarely mentioned by visitors of the period.

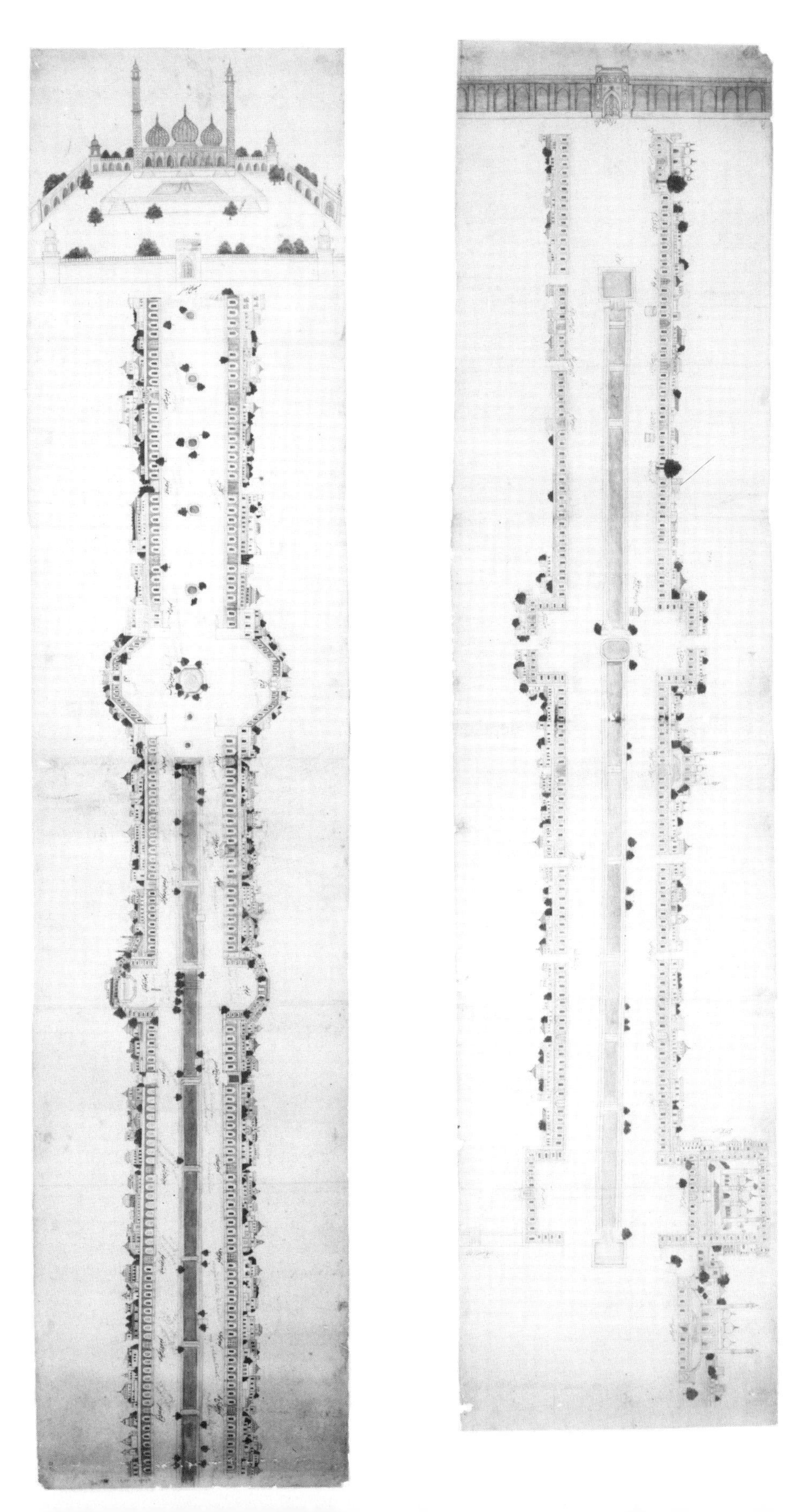

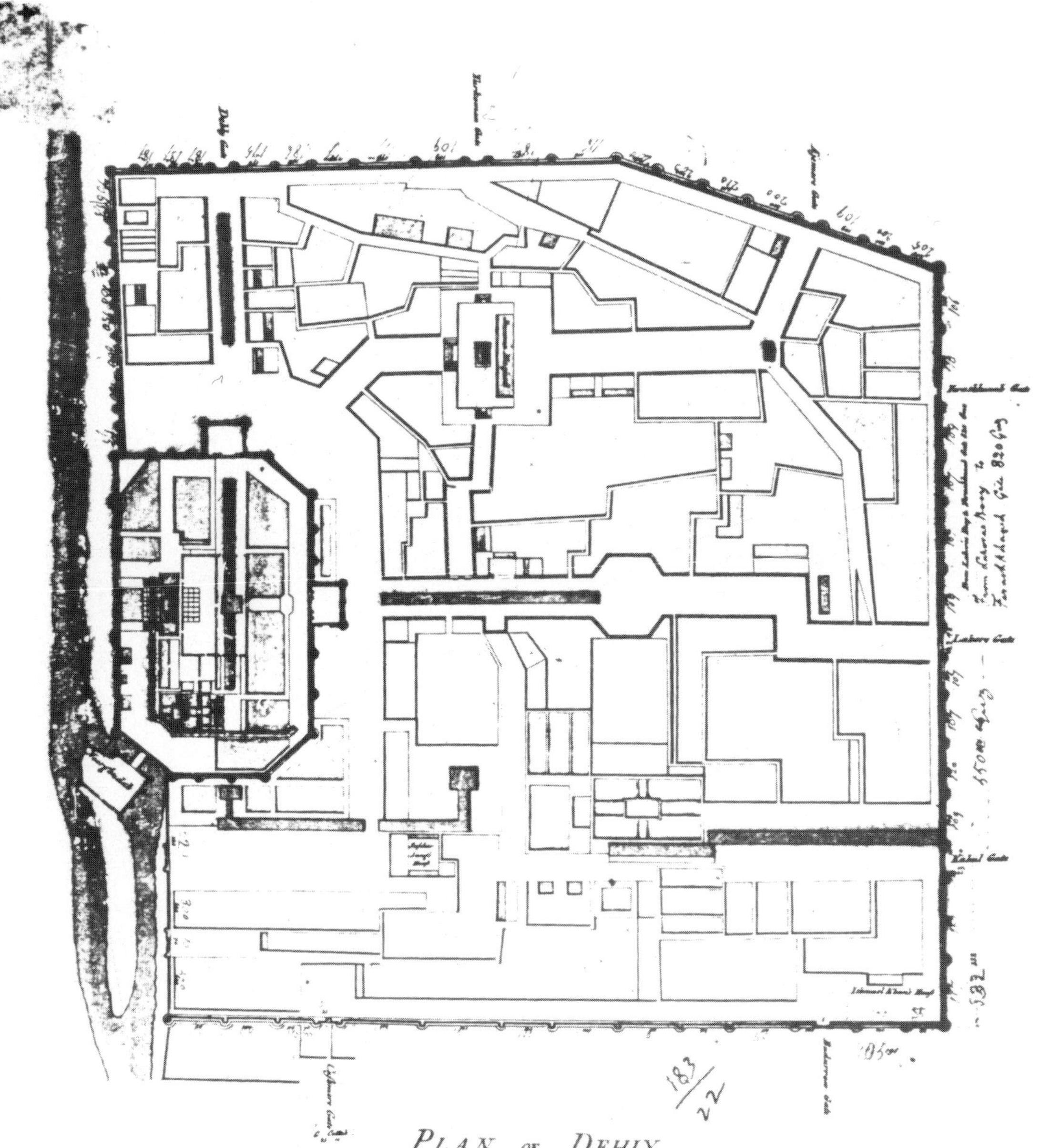

PLAN OF DEHLY
Reduced from a Large
HINDOSTANNY MAP of
That City

1800(?)

The location of Chowk Sa'ad-Ullah Khan appears to have been a problem to the early cartographers. In a map of Delhi in the British Library tentatively dated to 1857 [38], this large square is shown correctly south of the Delhi gate of the fort, with the Akbarabadi Masjid on the west side. The title reads: 'Cantonment & City, &c., of Delhi. Shewing the British Camp and Positions, as on the 1st August 1857, reduced from the official map and filled in by Capt. F.C. Maisey, D.J.A. Gent.' Handwritten numerals 1-38 and notes identify mainly military details, with 'subsequent additions' of ten more, and locations of Batteries and Picquets have been added with a stamp. So this is clearly a map to show the latest positions of the attacking forces. But in another map published by Edward Stanford in 1857 [39], this Chowk is shown north of the road from the fort to Jamma Masjid. The earliest edition of this map must also have been made during the seige, since an advertisement on the slip-cover announces 'a splendid chromo-lithograph of the city of Delhi' with special mention that 'This Print will become the more valuable and interesting from the fact that when the Arms of Britain shall have accomplished a just and terrible retribution, the City of Delhi will have ceased to exist.' Luckily this dire warning never came true, though the citizens who lived through the September rampages of the British soldiers must have thought their end had come. This displacement of the Sa'ad-Ullah Chowk was followed by Edward Weller in his decorative map [40] drawn for the *Weekly Dispatch* atlas the following year and seems to have remained on the maps even after the swathe of destruction around the fort ordered by the victorious British in the 1860s.

Conclusion

This survey of the maps and plans of Indian towns known to have been made before the middle of the 19th century has been brief. Other plans are known, and it is likely that more will come to light now that a search for them has been made. The finest of all is probably the very detailed plan of the walled town of Shahjahanabad in the India Office Libary that has now been fully translated for the first time. Since it was never published, it is difficult to assess its influence on later plans. The earliest British large-scale plan of Delhi was not made until 1868. This was a survey for the Punjab Revenue Survey, on eight sheets to a scale of 12 inches to the mile. Every lane is shown, but comparatively few of the buildings are named. It is interesting to see how far the

---

Tab. 2
A curious map of Delhi [32], known only from this English copy. The gates around the wall are named, (from upper left: Dehly, Turkoman, Ajemere, Farashkhanah, Lahore, Kabul, Budurrow, and Cashmere) and the distances in *gaz* between the bastions (the note on the left, induplicate, reads: From Lahoree Boory to Farashkhanah Gate 820 guz). Safdar Jang's house is named on a branch of the canal before it enters the fort, and that of Ishmael Khan at the lower right corner. Apart from the Jamma Masjid and Noorghudah (Salimgarh), there is no other text.

city had already spread beyond the old walls of Shahjahanabad. A corrected version was published five years later. What is now part of New Delhi west of the Mathura road is criss-crossed with 'Remains of former streets'. These must have been destroyed when New Delhi was laid out in the 20th century, yet one more of the many cities built on this site, that have gradually fallen into decay.

References

Maps referred to in the text in square brackets

[1] [Shatrunjaya pata]. Ahmedabad: Calico Museum of Textiles, Acc. 1043, size: 180 x 271 cm. Illustrated in Kay Talwar and Kalyan Krishna, Pigment Paintings on Cloth, Ahmedabad: Calico Museum of Textiles, 1979, p. 89; and Susan Gole: Indian Maps and Plans, New Delhi: Manohar, 1989, p. 56. Since most of the Indian plans are illustrated in this work it is hereafter referred to as simply IMAP.

[2] [Annakuta Pichhavai at Nathdwara]. Ahmedabad: Calico Museum of Textiles, Acc. 1561, size: 119 x 169 cm. Illustrated in Talwar and Krishna, op cit, pp. 36, 43; IMAP, p. 57-58.

[3] Two vijnaptapatras have been located: (i) Bikaner: Abhaya Jain Granth Bhandar, size: about 23 cm wide and some metres long; IMAP, p. 54; this scroll is an invitation to Udaipur. (ii) Baroda: Oriental Institute, M.S. University of Baroda, Cat. no. 7572, size: 23 cm x about 12 m; IMAP, pp. 54, 55; an invitation to Jaiselmer.

[4] [Vraj Yatra Pichhavai]. Ahmedabad: Calico Museum of Textiles, Acc. 1330, size: 180 x 193 cm. Talwar and Krishna, op cit, pp. 26-27; IMAP, p. 61. Also a modern plan, Vraj Chaurassi Kos, Baroda: private collection, size: 30 x 49 cm, IMAP, p. 61.

[5] [Jagannath temple, Orissa]. Paris: Bibliothèque Nationale, Suppl. Persan 1041, size: 270 x 150 cm; IMAP, p. 63.

[6] [Ujjain] New Delhi: National Museum, Acc. 59.1284/7, size: 32.5 x 24 cm; IMAP, p. 64.

[7] [Plan of Kashi]. Baroda: private collection, size: 50 x 64 cm; IMAP, p. 65.

[8] Kashi. Paris: private collection, size: 72 x 76 cm; IMAP, p. 66.

[9] [Varanasi]. New Delhi: National Museum, Cat. no. 61.935, size: 325 x 229 cm; IMAP, p. 65.

[10] Sankhoder Bet. Bombay: Prince of Wales Museum, Acc. 70.4, size: 32 x 25 cm; IMAP, p. 69.

[11] [Dwarka]. Jaipur: City Palace Museum, Cat. no. 139, size: 178 x 175 cm; Chandramani Singh in Skelton, R.: Facets of Indian Art, London: Victoria & Albert Museum, 1986, pp. 186-88, IMAP, p. 70.

[12] This temple and town have not been identified, though they seem to be very precisely drawn. The lay-out of streets in relation to the temple fits the island on the Caveri at Srirangam, but the famous temple there is dedicated to Vishnu, and in the painting is a Shaivite temple. New Delhi: private collection, size: 63.5 x 122 cm, IMAP, p. 67.

[13] At Amber (Rajasthan): Bhojan Shala. IMAP, pp. 171-73.

[14] [Surat]. Jaipur: City Palace Museum, Cat. no. 118, size: 186 x 210 cm; Chandramani Singh in Facets of Indian Art, pp. 190-91. IMAP, pp. 164-65.

[15] [Agra]. Jaipur: City Palace Museum, Cat. no. 126, size: 292 x 272 cm; Chandramani Singh, op cit, pp. 190-92, IMAP, pp. 200-01.

[16] [Amber]. New Delhi: National Museum, Cat. no. 56.92.4; IMAP, pp. 170-71. This map was formerly with the Jaipur collection.

[17] [Surat]. From Anis al-Haj, Bombay: Prince of Wales Museum, size: c.24 x 40 cm; S. Gorakshkar in Facets of Indian Art, pp. 158-67, IMAP, p. 162.

[18] [Kashmir]. Jaipur: City Palace Museum, Cat. no. 120, size: 280 x 223 cm; Chandramani Singh, op cit, pp. 189-90, IMAP, pp. 116-17.

[19] [Jammu]. New Delhi: National Museum, Cat. no. 58.33/4, size: 198 x 121 cm; IMAP, p. 166.

[20] [Nasik]. Pune: Peshwe Daftar, size: 300 x 315 cm; IMAP, pp. 168-69.

[21] Three maps of Bijapur are known. The original appears to be the one at Bijapur in the Archaeological Museum, Gol Gumbaz, size: 102 x 149 cm; two others are at Hyderabad in the Andhra Pradesh State Archives, one (71 x 90 cm) very similar to that at Bijapur, and a smaller one (43 x 60 cm) more stylised; IMAP, 160-61.

[22] Udgir. Hyderabad: Andhra Pradesh State Archives, size: circular 67 cm diam.; IMAP, p. 167.

[23] [Dharwar Fort]. Pune: Deccan College Museum, size: irregular 110 x 80 cm; K.N. Chitnis in Studies in Indian History and Culture, Dharwar: Karnatak Education Board, nd; IMAP, p. 148.

[24] [Janjira]. Pune: Raja Dinkar Kelkar Museum, size: 95 x 70 cm; S.R. Tikekar in The Illustrated Weekly of India, 1949, March 20, IMAP, p. 153.

[25] Vijaydurg. Bombay: Prince of Wales Museum, Cat. no. 55.103, size: 190.5 x 172.5 cm; B.K. Apte in A History of the Maratha Navy, Bombay: State Board for Literature and Culture, 1973, Fig. 7; G.D. Deshpande in The Indian Archives, 1953, pp. 88-89; IMAP, p. 137.

[26] [Hyderabad]. Hyderabad: Idara Adabiyat-e-Urdu, size: 275 x 215 cm; IMAP, p. 190 (part).

[27] [Lal Qila, Shahjahanabad]. Jaipur: City Palace Museum, Cat. no. 122, size: 137 x 64 cm; IMAP, p. 176.

[28] [Agra Fort]. Jaipur: City Palace Museum, Cat. no. 125, size: 121 x 83 cm; IMAP, p. 175.

[29] [Shahjahanabad: Fort, Chandni Chowk and Faiz Bazar]. London: Victoria & Albert Museum, Cat. nos AL 1763, AL 1762, AL 1754, sizes: 75 x 82 cm, 31 x 140 cm, 31 x 135 cm; S. Gole in South Asian Studies (Cambridge), 1988, pp. 13-27, IMAP, pp. 178-79.

[30] [Ali Mardan Khan Canal]. Hyderabad: Andhra Pradesh State Archives, size 43 x 1250 cm; IMAP, pp. 104-09.

[31] [Shahjahanabad to Kandahar]. London: India Office Library & Records, Cat. nos Pers. Mss 4725 and 4380, sizes: approx. 25 x 2000 cm, 20 x 1200 cm. IMAP, pp. 94-103.

[32] Plan of Dehly. New Delhi: National Archives, Cat. no. F.183/22, size: c.35 x 35 cm; IMAP, p. 174.

[33] Trigonometircal Survey of the Environs of Delhy or Shah Jehanabad. 1808. London: India Office Libary & Records, Cat. no. E.VII.16, size: 61 x 46 cm.

[34] Plan of Delhi. Copied and Zincographed at the Ordnance Survey Office, Southampton, from a M.S. plan in the possession of Col. Edward Harvey. Unattached. August 3rd 1857. London: India Office Libary & Records, Cat. no. E.VII.20, size: 71 x 84 cm.

[35] (i) Map of the District of Dihlee Surveyed by Capts T. Oliver, L.H. Simmonds and W. Brown, Revenue Surveyors in 1824-25-26-30-31-32 and 40. London: India Office Libarary & Records, Cat. no. E.VII.14, size: 63.5 x 51 cm; (ii) map of the District of Dihlee Surveyed by Captains T. Oliver, I.H. Simmonds... reduced and Lithographed in the Supt of Revenue Surveyor's Office under the orders of H.M. Elliot Esq., Sudder Board of Revenue, NWP. London: India Office Libary & Records, Cat. no. E.VII.13, size: 53 x 56 cm.

[36] A Topographical Survey of a portion of the Dihlee District shewing the principal Basins of Drainage..., Surveyed by Lieut. F.J. Burgess, 14th Beng N.I., Assistant Revenue Surveyor in 1849-50. London: Private collection, size: 1470 x 2170 in 2 sheets.

[37] The Fort and Cantonment of Delhi. London: India Office Library & Records, Cat. nos. E.VII.18 and E.VII.19, sizes: 56 x 38 cm; 61 x 51 cm.

[38] Cantonment & City &c of Delhi...London: British Library Map Room, Cat. no. 55455. (11), size: 49 x 42 cm.

[39] Delhi and its Environs. From plan and other original materials, transmitted from India, and the surveys of the Hon. East India Company. Published London: Edward Stanford, 6 Charing Cross. Maclure, Macdonald & Macgregor, 57 Walbrook, London. Lithographers by Steam Power. London: British Library Map Room, Cat. no. 55455. (2); size: 35.5 x 48.5 cm.

[40] Plan of Delhi and its Environs. 1857. Drawn by Edward Weller and published in the Weekly Dispatch. London: private collection, size: 44 x 31 cm.

[41] Cantonment, City & Environs of Delhi. 1867-68. Surveyed under the Superintendence of Colonel J.E. Gastrell...Punjab Revenue Survey. London: British Libary Map Room. 8 sheets, each 57.5 x 80 cm.

25. **Fritz Dörrenhaus: Urbanität und gentile Lebensform.** Der europäische Dualismus mediteraner und indoeuropäischer Verhaltensweisen, entwickelt aus einer Diskussion um den Tiroler Einzelhof. 1970. 64 S., 5 Ktn., kt. **ISBN 3-515-00532 - 3**
26. **Eckart Ehlers / Fred Scholz / Günter Schweizer: Strukturwandlungen im nomadisch-bäuerlichen Lebensraum des Orients.** Eckart Ehlers: Turkmenensteppe. Fred Scholz: Belutschistan. Günter Schweizer: Azerbaidschan. 1970. VI, 148 S. m. 4 Abb., 4 Taf., 20 Ktn., kt. **2228 - 7**
27. **Ulrich Schweinfurth / Heidrun Marby / Klaus Weitzel / Klaus Hausherr / Manfred Domrös: Landschaftsökologische Forschungen auf Ceylon.** 1971. VI, 232 S. m. 46 Abb., 10 Taf. m. 20 Bildern, 1 Falttaf., kt. (vgl. Bd. 54) **0533 - 1**
28. **Georges Henri Lutz: Republik Elfenbeinküste.** 1971. VI, 48 S. m. 7 Ktn. u. 2 Abb., kt. **0534 - X**
29. **Harry Stein: Die Geographie an der Universität Jena (1786-1939).** Ein Beitrag zur Entwicklung der Geographie als Wissenschaft. Vorgelegt von **Joachim H. Schultze.** 1972. XII, 152 S., 16 Taf. m. 4 Ktn. u. 19 Abb., kt. **0535 - 8**
30. **Arno Semmel: Geomorphologie der Bundesrepublik Deutschland.** Grundzüge, Forschungsstand, aktuelle Fragen - erörtert an ausgewählten Landschaften. 4., völlig überarbeitete u. erw. Aufl. 1984. 192 S. m. 57 Abb., kt. **4217-2**
31. **Hermann Hambloch: Allgemeine Anthropogeographie.** Eine Einführung. 5., neubearb. Aufl. 1982. XIII, 268 S. m. 40 Abb. (davon 16 Faltktn.), 37 Tab., 12 Fig., kt. **3618 - 0**
32. **Arno Semmel,** Hrsg.: **Neue Ergebnisse der Karstforschung in den Tropen und im Mittelmeerraum.** Vorträge des Frankfurter Karstsymposiums. Zusammengestellt von **Karl-Heinz Pfeffer.** 1973. XX, 156 S. m. 35 Abb. u. 63 Bildern, kt. **0538 - 2**
33. **Emil Meynen,** Hrsg.: **Geographie heute - Einheit und Vielfalt. Ernst Plewe** zu seinem 65. Geburtstag von Freunden und Schülern gewidmet. Hrsg. unter Mitarbeit von **Egon Riffel.** 1973. X, 425 S. m. 39 Abb., 26 Bildern u. 14 Ktn., kt. **0539 - 0**
34. **Jürgen Dahlke: Der Weizengürtel in Südwestaustralien.** Anbau und Siedlung an der Trockengrenze. 1973. XII, 275 S., 67 Abb., 4 Faltktn., kt. **0540 - 4**
35. **Helmut J. Jusatz,** Hrsg.: **Fortschritte der geomedizinischen Forschung.** Beiträge zur Geoökologie der Infektionskrankheiten. Vorträge d. Geomedizin. Symposiums auf Schloß Reisenburg v. 8.-12. Okt. 1972. Herausgegeben im Auftrag der Heidelberger Akademie der Wissenschaften. 1974. VIII, 164 S. m. 47 Abb., 8 Bildern u. 2 Falttaf., kt. **1797 - 6**
36. **Werner Rutz,** Hrsg.: **Ostafrika** - Themen zur wirtschaftlichen Entwicklung am Beginn der Siebziger Jahre. Festschrift **Ernst Weigt.** 1974. VIII, 176 S. m. 17 Ktn., 7 Bildern u. 1 Abb., kt. **1796 - 8**
37. **Wolfgang Brücher: Die Industrie im Limousin.** Ihre Entwicklung und Förderung in einem Problemgebiet Zentralfrankreichs. 1974. VI, 45 S. m. 10 Abb. u. 1 Faltkte., kt. **1853 - 0**
38. **Bernd Andreae: Die Farmwirtschaft an den agronomischen Trockengrenzen.** Über den Wettbewerb ökologischer Varianten in der ökonomischen Evolution. Betriebs- und standortsökonomische Studien in der Farmzone des südlichen Afrika und der westlichen USA. 1974. X, 69 S.,m. 14 Schaubildern u. 24 Übersichten, kt. **1821 - 2**
39. **Hans-Wilhelm Windhorst: Studien zur Waldwirtschaftsgeographie.** Das Ertragspotentialder Wälder der Erde. Wald- und Forstwirtschaft in Afrika. Ein forstgeographischer Überblick. 1974. VIII, 75 S. m. 10 Abb., 8 Ktn., 41 Tab., kt. **2044 - 6**
40. **Hilgard O'Reilly Sternberg: The Amazon River of Brazil.** (vergriffen) **2075 - 6**
41. **Utz Ingo Küpper / Eike W. Schamp,** Hrsg.: **Der Wirtschaftsraum.** Beiträge zur Methode und Anwendung eines geographischen Forschungsansatzes. Festschrift für **Erich Otremba** zu seinem 65. Geburtstag. 1975. VI, 294 S. m. 10 Abb., 15 Ktn., kt. **2156 - 6**
42. **Wilhelm Lauer,** Hrsg.: **Landflucht und Verstädterung in Chile.** Exodu rura yl urbanización en Chile. Mit Beiträgen von Jürgen Bähr, Winfried Golte und Wilhelm Lauer. 1976. XVIII, 149 S., 13 Taf. m. 25. Fotos. 41 Figuren, 3 Faltktn., kt. **2159 - 0**
43. **Helmut J. Jusatz,** Hrsg.: **Methoden und Modelle der geomedizinischen Forschung.** Vorträge des 2. Geomedizin. Symposiums auf Schloß Reisensburg vom 20.-24. Okt. 1974. Hrsg. im Auftrag der Heidelberger Akademie der Wissenschaften. 1976. X, 174 S. m. 7 Abb., 2 Diagr., 20 Tab., 24 Ktn., Summaries, 6 Taf. m. 6 Bildern, kt. **2308 - 9**
44. **Fritz Dörrenhaus: Villa und Villegiatura in der Toskana.** Eine italienische Institution und ihre gesellschaftsgeographische Bedeutung. Mit einer einleitenden Schilderung "Toskanische Landschaft" von Herbert Lehmann. 1976. X, 153 S. m. 5 Ktn., 1 Abb., 1 Schema (Beilage), 8 Taf. m. 24 Fotos, 14 Zeichnungen von Gino Canessa, Florenz, u. 2 Stichen, kt. **2400 - X**
45. **Hans Karl Barth: Probleme der Wasserversorgung in Saudi-Arabien.** 1976. VI, 33 S. m. 3 Abb., 4 Tab., 4 Faltktn., 1 Kte., kt. **2401 - 8**
46. **Hans Becker / Volker Höhfeld / Horst Kopp: Kaffee aus Arabien.** Der Bedeutungswandel eines Weltwirtschaftsgutes und seine siedlungsgeographische Konsequenz an der Trockengrenze der Ökumene. 1979. VIII, 78 S. m. 6 Abb., 6 Taf. m. 12 Fotos, 2 Faltktn. kt. **2881 - 1**
47. **Hermann Lautensach: Madeira, Ischia und Taormina.** Inselstudien. 1977. XII, 57 S. m. 16 Abb., 5 Ktn., kt. **2564 - 2**
48. **Felix Monheim: 20 Jahre Indianerkolonisation in Ostbolivien.** 1977. VI, 99 S., 14 Ktn., 17 Tab., kt. **2563 - 4**
49. **Wilhelm Müller-Wille: Stadt und Umland im südlichen Sowjet-Mittelasien.** 1978. VI, 48 S. m. 20 Abb. u. 7 Tab., kt. **2762 - 9**
50. **Ernst Plewe,** Hrsg.: **Die Carl Ritter-Bibliothek.** Nachdruck der Ausg. Leipzig, Weigel, 1861: "Verzeichnis der Bibliothek und Kartensammlung des Professors, Ritters etc. etc. Doktor Carl Ritter in Berlin." 1978. XXVI, 565 S., Frontispiz, kt. **2854 - 4**

51. **Helmut J. Jusatz,** Hrsg.: **Geomedizin in Forschung und Lehre.** Beiträge zur Geoökologie des Menschen. Vorträge des 3. Geomed. Symposiums auf Schloß Reisensburg vom 16. - 20. Okt. 1977. Hrsg. im Auftrag der Heidelberger Akademie der Wissenschaften. 1979. XV, 122 S. m. 15 Abb. u. 14 Tab., 1 Faltkte., Summaries, kt. **2801 - 3**
52. **Werner Kreuer: Ankole.** Bevölkerung - Siedlung - Wirtschaft eines Entwicklungsraumes in Uganda. 1979. XI, 106 S. m. 11 Abb., 1 Luftbild auf Falttaf., 8 Ktn., 18 Tab., kt. **3063 - 8**
53. **Martin Born: Siedlungsgenese und Kulturlandschaftsentwicklung in Mitteleuropa.** Gesammelte Beiträge. Hrsg. im Auftrag des Zentralausschusses für Deutsche Landeskunde von **Klaus Fehn.** 1980. XL, 528 S. m. 17 Abb., 39 Ktn. kt. **3306 - 8**
54. **Ulrich Schweinfurth / Ernst Schmidt-Kraepelin / Hans Jürgen von Lengerke / Heidrun Schweinfurth-Marby / Thomas Gläser / Heinz Bechert:** Forschungen **auf Ceylon II.** 1981. VI, 216 S. m. 72 Abb., kt. (Bd. I s. Nr. 27) **3372 - 6**
55. **Felix Monheim: Die Entwicklung der peruanischen Agrarreform 1969-1979 und ihre Durchführung im Departement Puno.** 1981. V, 37 S. m. 15 Tab., kt. **3629 - 6**
56. **- / Gerrit Köster: Die wirtschaftliche Erschließung des Departement Santa Cruz (Bolivien) seit der Mitte des 20. Jahrhunderts.** 1982. VIII, 152 S. m. 2 Abb. u. 12 Ktn., kt. **3635 - 0**
57. **Hans Georg Bohle: Bewässerung und Gesellschaft im Cauvery-Delta (Südindien).** Eine geographische Untersuchung über historische Grundlagen und jüngere Ausprägung struktureller Unterentwicklung. 1981. XVI, 266 S. m. 33 Abb., 49 Tab., 8 Kartenbeilagen, kt. **3550 - 8**
58. **Emil Meynen / Ernst Plewe,** Hrsg.: **Forschungsbeiträge zur Landeskunde Süd- und Südostasiens.** Festschrift für **Harald Uhlig** zu seinem 60. Geburtstag, Band 1. 1982. XVI, 253 S. m. 45 Abb. u. 11 Ktn., kt. **3743 - 8**
59. **- / -,** Hrsg.: **Beiträge zur Hochgebirgsforschung und zur Allgemeinen Geographie.** Festschrift für **Harald Uhlig** zu seinem 60. Geburtstag, Band 2. 1982. VI, 313 S. m. 51 Abb. u. 6 Ktn., 1farb. Faltkte., kt. **3744 - 6**
Beide Bände zus. kt. **3779 - 9**
60. **Gottfried Pfeifer: Kulturgeographie in Methode und Lehre.** Das Verhältnis zu Raum und Zeit. Gesammelte Beiträge. 1982. XI, 471 S. m. 3 Taf., 18 Fig., 16 Ktn., 15 Tab. u. 7 Diagr., kt. **3668 - 7**
61. **Walter Sperling: Formen, Typen und Genese des Platzdorfes in den böhmischen Ländern.** Beiträge zur Siedlungsgeographie Ostmitteleuropas. 1982. X, 187 S. m. 39 Abb., kt. **3654 - 7**
62. **Angelika Sievers: Der Tourismus in Sri Lanka (Ceylon).** Ein sozialgeographischer Beitrag zum Tourismusphänomen in tropischen Entwicklungsländern, insbesondere in Südasien. 1983. X, 138 S. m. 25 Abb. u. 19 Tab., kt. **3889 - 2**
63. **Anneliese Krenzlin: Beiträge zur Kulturlandschaftsgenese in Mitteleuropa.** Gesammelte Aufsätze aus vier Jahrzehnten, hrsg. von **H.-J. Nitz** u. **H. Quirin.** 1983. XXXVIII, 366 S. m. 55 Abb., kt. **4035 - 8**
64. **Gerhard Engelmann: Die Hochschulgeographie in Preußen 1810-1914.** 1983. XII, 184 S., 4 Taf., kt. **3984 - 8**
65. **Bruno Fautz: Agrarlandschaften in Queensland.** 1984. 195 S. m. 33 Ktn., kt. **3890 - 6**
66. **Elmar Sabelberg: Regionale Stadttypen in Italien.** Genese und heutige Struktur der toskanischen und sizilianischen Städte an den Beispielen Florenz, Siena, Catania und Agrigent. 1984. XI, 211 S. m. 26 Tab., 4 Abb., 57 Ktn. u. 5 Faltktn., 10 Bilder auf 5 Taf., kt. **4052 - 8**
67. **Wolfhard Symader: Raumzeitliches Verhalten gelöster und suspendierter Schwermetalle.** Eine Untersuchung zum Stofftransport in Gewässern der Nordeifel und niederrheinischen Bucht. 1984. VIII, 174 S. m. 67 Abb., kt. **3909 - 0**
68. **Werner Kreisel: Die ethnischen Gruppen der Hawaii-Inseln.** Ihre Entwicklung und Bedeutung für Wirtschaftsstruktur und Kulturlandschaft. 1984. X, 462 S. m. 177 Abb. u. 81 Tab., 8 Taf. m. 24 Fotos, kt. **3412 - 9**
69. **Eckart Ehlers: Die agraren Siedlungsgrenzen der Erde.** Gedanken zur ihrer Genese und Typologie am Beispiel des kanadischen Waldlandes. 1984. 82 S. m. 15 Abb., 2 Faltktn., kt. **4211 - 3**
70. **Helmut J. Jusatz / Hella Wellmer,** Hrsg.: **Theorie und Praxis der medizinischen Geographie und Geomedizin.** Vorträge der Arbeitskreissitzung Medizinische Geographie und Geomedizin auf dem 44. Deutschen Geographentag in Münster 1983. Hrsg. im Auftrage des Arbeitskreises. 1984. 85 S. m. 20 Abb., 4 Fotos u. 2 Kartenbeilagen, kt. **4092 - 7**
71. **Leo Waibel †: Als Forscher und Planer in Brasilien:** Vier Beiträge aus der Forschungstätigkeit 1947-1950 in Übersetzung. Hrsg. von **Gottfried Pfeiffer** u. **Gerd Kohlhepp.** 1984. 124 S. m. 5 Abb., 1 Taf., kt. **4137 - 0**
72. **Heinz Ellenberg: Bäuerliche Bauweisen in geoökologischer und genetischer Sicht.** 1984. V, 69 S. m. 18 Abb., kt. **4208 - 3**
73. **Herbert Louis: Landeskunde der Türkei.** Vornehmlich aufgrund eigener Reisen. 1985. XIV, 268 S. m. 4 Farbktn. u. 1 Übersichtskärtchen des Verf., kt. **4312 - 8**
74. **Ernst Plewe / Ute Wardenga: Der junge Alfred Hettner.** Studien zur Entwicklung der wissenschaftlichen Persönlichkeit als Geograph, Länderkundler und Forschungsreisender. 1985. 80 S. m. 2 Ktn. u. 1 Abb., kt. **4421 - 3**
75. **Ulrich Ante: Zur Grundlegung des Gegenstandsbereiches der Politischen Geographie.** Über das "Politische" in der Geographie. 1985. 184 S., kt. **4361 - 6**
76. **Günter Heinritz / Elisabeth Lichtenberger,** eds.: **The Take-off of Suburbia and the Crisis of the Central City.** Proceedings of the International Symposium in Munich and Vienna 1984. 1986. X, 300 S. m. 95 Abb., 49 Tab., kt. **4402 - 7**
77. **Klaus Frantz: Die Großstadt Angloamerikas im Wandel des 18. und 19. Jahrhunderts.** Versuch einer sozialgeographischen Strukturanalyse anhand ausgewählter Beispiele der Nordostküste. 1987. 200 S. m. 32 Ktn. u. 12 Abb. kt. **4433 - 7**
78. **Claudia Erdmann: Aachen im Jahre 1812.** Wirtschafts- und sozialräumliche Differenzierung einer frühindustriellen Stadt. 1986. VIII, 257 S. m. 6 Abb., 44 Tab., 19 Fig., 80 Ktn., kt. **4634 - 8**
79. **Josef Schmithüsen †: Die natürliche Lebewelt Mitteleuropas.** Hrsg. von **Emil Meynen.** 1986. 71 S. m. 1 Taf., kt. **4638 - 8**
80. **Ulrich Helmert: Der Jahresgang der Humidität in Hessen und den angrenzenden Gebieten.** 1986. 108 S. m. 11 Abb. u. 37 Ktn. i. Anh., kt. **4630 - 5**
81. **Peter Schöller: Städtepolitik, Stadtumbau und Stadterhaltung in der DDR.** 1986. 55 S., 4 Taf. m. 8 Fotos, 12 Ktn., kt. **4703 - 4**

82. **Hans-Georg Bohle: Südindische Wochenmarktsysteme.** Theoriegeleitete Fallstudien zur Geschichte und Struktur polarisierter Wirtschaftskreisläufe im ländlichen Raum der Dritten Welt. 1986. XIX, 291 S. m. 43 Abb., 12 Taf., kt. **4601 - 1**
83. **Herbert Lehmann: Essays zur Physiognomie der Landschaft.** Mit einer Einleitung von Renate Müller, hrsg. von **Anneliese Krenzlin** und **Renate Müller.** 1986. 267 S. m. 25 s/w- und 12 Farbtaf., kt. **4689 - 5**
84. **Günther Glebe / J. O'Loughlin,** eds.: **Foreign Minorities in Continental European Cities.**1987. 296 S. m. zahlr. Ktn. u. Fig., kt. **4594 - 5**
85. **Ernst Plewe †: Geographie in Vergangenheit und Gegenwart.** Ausgewählte Beiträge zur Geschichte und Methode des Faches. Hrsg. von **Emil Meynen** und **Uwe Wardenga.** 1986. 438 S., kt. **4791 - 3**
86. **Herbert Lehmann †: Beiträge zur Karstmorphologie.** Hrsg. von **F. Fuchs, A. Gerstenhauer, K.-H. Pfeffer.** 1987. 251 S. m. 60 Abb., 2 Ktn., 94 Fotos, kt. **4897 - 9**
87. **Karl Eckart: Die Eisen- und Stahlindustrie in den beiden deutschen Staaten.** 1988. 277 S. m. 167 Abb., 54 Tab., 7 Übers., kt. **4958 - 4**
88. **Helmut Blume / Herbert Wilhelmy,** Hrsg.: **Heinrich Schmitthenner Gedächtnisschrift.** Zu seinem 100. Geburtstag. 1987. 173 S. m. 42 Abb., 8 Taf., kt.**5033 - 7**
89. **Benno Werlen: Gesellschaft, Handlung und Raum** (vergriffen, 2., durchges. Aufl. 1988 s.S. 180) **4886-3**
90. **Rüdiger Mäckel / Wolf-Dieter Sick,** Hrsg.: **Natürliche Ressourcen und ländliche Entwicklungsprobleme der Tropen.** Festschrift für **Walther Manshard.** 1988. 334 S. m. zahlr. Abb., kt. **5188-0**
91. **Gerhard Engelmann †: Ferdinand von Richthofen 1833–1905. Albrecht Penck 1858–1945.** Zwei markante Geographen Berlins. Aus dem Nachlaß hrsg. von **Emil Meynen.** 1988. 37 S. m. 2 Abb., kt. **5132-5**
92. **Gerhard Hard: Selbstmord und Wetter – Selbstmord und Gesellschaft.** Studien zur Problemwahrnehmung in der Wissenschaft und zur Geschichte der Geographie. 1988. 356 S., 11 Abb., 13 Tab., kt. **5046-9**
93. **Siegfried Gerlach: Das Warenhaus in Deutschland.** Seine Entwicklung bis zum Ersten Weltkrieg in historisch-geographischer Sicht. 1988. 178 S. m. 33 Abb., kt. **5103-1**
94. **Walter H. Thomi: Struktur und Funktion des produzierenden Kleingewerbes in Klein- und Mittelstädten Ghanas.** Ein empirischer Beitrag zur Theorie der urbanen Reproduktion in Ländern der Dritten Welt. 1989. XVI, 312 S., kt. **5090-6**
95. **Thomas Heymann: Komplexität und Kontextualität des Sozialraumes.** 1989. VIII, 511 S. m. 187 Abb., kt. **5315-8**
96. **Dietrich Denecke / Klaus Fehn,** Hrsg.: **Geographie in der Geschichte.** (Vorträge der Sektion 13 des Deutschen Historikertags, Trier 1986.) 1989. 97 S. m. 3 Abb., kt. DM 36,– **5428-6**
97. **Ulrich Schweinfurth,** Hrsg.: **Forschungen auf Ceylon III.** Mit Beiträgen von C. Preu, W. Werner, W. Erdelen, S. Dicke, H. Wellmer, M. Bührlein u. R. Wagner. 1989.258 S. m. 76 Abb., kt. **5084-1**
98. **Martin Boesch: Engagierte Geographie.** 1989. XII, 284 S., kt. **5514-2**
99. **Hans Gebhardt: Industrie im Alpenraum.** Alpine Wirtschaftsentwicklung zwischen Außenorientierung und endogenem Potential. 1990. 283 S. m. 68 Abb., kt. **5397-2**
100. In Vorbereitung
101. **Siegfried Gerlach: Die deutsche Stadt des Absolutismus im Spiegel barocker Veduten und zeitgenössischer Pläne.** Erweiterte Fassung eines Vortrags am 11. November 1986 im Reutlinger Spitalhof. 1990. 80 S. m. 32 Abb., dav. 7 farb., kt. **5600-9**
102. **Peter Weichhart: Raumbezogene Identität.** Bausteine zu einer Theorie räumlich-sozialer Kognition und Identifikation. 1990. 118 S., kt. **5701-3**
103. **Manfred Schneider: Beiträge zur Wirtschaftsstruktur und Wirtschaftsentwicklung Persiens 1850-1900.** Binnenwirtschaft und Exporthandel in Abhängigkeit von Verkehrserschließung, Nachrichtenverbindungen, Wirtschaftsgeist und politischen Verhältnissen anhand britischer Archivquellen. 1990. XII, 381 S. m. 86 Tab., 16 Abb., kt. **5458-8**
104. **Ulrike Sailer-Fliege: Der Wohnungsmarkt der Sozialmietwohnungen.** Angebots- und Nutzerstrukturen dargestellt an Beispielen aus Nordrhein-Westfalen. 1991. XII, 287 S. m. 92 Abb., 30 Tab., 6 Ktn., kt.**5836-2**
105. **Helmut Brückner / Ulrich Radtke,** Hrsg.: **Von der Nordsee bis zum Indischen Ozean/From the North Sea to the Indian Ocean.** Ergebnisse der 8. Jahrestagung des Arbeitskreises „Geographie der Meere und Küsten", 13.-15. Juni 1990, Düsseldorf / Results of the 8th Annual Meeting of the Working group „Marine and Coastal Geography", June 13-15, 1990, Düsseldorf. 1991. 264 S. mit 117 Abbildungen, 25 Tabellen, kt. **5898-2**
106. **Heinrich Pachner: Vermarktung landwirtschaftlicher Erzeugnisse in Baden-Württemberg.** 1992. 238 S. m. 53 Tab., 15 Abb. u. 24 Ktn., kt. **5825-7**
107. **Wolfgang Aschauer: Zur Produktion und Reproduktion einer Nationalität – die Ungarndeutschen.** 1992. 315 S. m. 85 Tab., 8 Ktn., 9 Abb., kt. **6082-0**
108. **Hans-Georg Möller: Tourismus und Regionalentwicklung im mediterranen Südfrankreich.** Sektorale und regionale Entwicklungseffekte des Tourismus - ihre Möglichkeiten und Grenzen am Beispiel von Côte d'Azur, Provence und Languedoc-Roussillon. 1992. XIV, 413 S. m. 60 Abb., kt. **5632-7**
109. **Klaus Frantz: Die Indianerreservationen in den USA.** Aspekte der territorialen Entwicklung und des sozioökonomischen Wandels. 1993. 298 S. m. 20 Taf., kt., **6217-3**
110. **Hans-Jürgen Nitz,** ed.: **The Early Modern World-System in Geographical Perspective.** 1993. XII, 403 S. m. 67 Abb., kt. **6094-4**
111. **Eckart Ehlers/Thomas Krafft,** Hrsg.: **Shâhjahânâbâd/Old Delhi.** Islamic Tradition and Colonial Change. 1993. 106 S. m. 14 Abb., 1 mehrfbg. Faltkt., 1 fbg. Frontispiz, kt. **6218-1**
112. **Ulrich Schweinfurth,** Hrsg.: **Neue Forschungen im Himalaya.** 1993. 293 S. m. 6 Ktn., 50 Abb., 35 Photos u. 1 Diagr., kt. **6263-7**

FRANZ STEINER VERLAG STUTTGART

ISSN 0425 - 1741